QUANTITATIVE ECOLOGICAL THEORY:

An Introduction to Basic Models

This book is concerned with the mathematical basics of ecological theory. It introduces the reader to the construction and analysis of models prevalent in current theory. The book is not intended as a general text, but more as a manual, with relevant background information, on how to construct and develop models in a mathematical way. Of the major theoretical 'issues' in ecology, the problems of ecosystem stability, co-existence of competitors, chaos, predator-prey cycles, and multiple stable-states are all considered, as they arise naturally from discussion of particular models. Some basic grounding in ecology, population biology, and mathematics is assumed on the part of readers, who are assumed to be mainly students taking advanced courses in ecology.

Michael R. Rose, Department of Biology, Dalhousie University, Canada

QUANTITATIVE ECOLOGICAL THEORY

AN INTRODUCTION TO BASIC MODELS

Michael R. Rose

CROOM HELM
London & Sydney

© 1987 Michael R. Rose
Croom Helm Ltd, Provident House, Burrell Row,
Beckenham, Kent, BR3 1AT
Croom Helm Australia, 44-50 Waterloo Road,
North Ryde, 2113, New South Wales

British Library Cataloguing in Publication Data

Rose, Michael R.
    Quantitative ecological theory: an
    introduction to basic models.
    1. Ecology — Mathematical models
    I. Title
    574.5′0724   QH541.15.M3

ISBN 0-7099-2288-4
ISBN 0-7099-2289-2

Printed and bound in Great Britain
by Billing & Sons Limited, Worcester.

CONTENTS

TO MY PARENTS

PREFACE

This is an inadvertent book, though it did arise
naturally enough from a course I give in
theoretical ecology. But I wouldn't have given the
course at all if one colleague in my department
hadn't left for a leave of absence, while another
abruptly resigned. This propelled me to the fore
where this teaching responsibility was concerned,
one I had never had any intention of discharging.

Then it turned out that one of my students was
regularly unable to make half the classes. As a
result, I began giving him my lecture notes each
week. As I knew that someone else would be
reading them, I began to write my notes more
carefully. Naturally enough, the other students
soon began to demand the notes too. Eventually
they were indulged. Thus I found myself writing a
textbook manuscript. By the next year, the
students were handed all their notes in one package
at the outset. But these were still just
hand-written. Inevitably, the demand that they be
typed arose.

This I didn't want to do until I found a
publisher. As it turned out, Tim Hardwick of Croom
Helm was willing to have his firm fill this role,
to my great satisfaction . . . and his
considerable frustration. I have been a desultory
author about producing this final text, and can
only express my gratitude for his enduring patience
over more than 18 months of delays.

I have a number of people to thank other than
Tim. Firstly, of course, there are my theoretical
ecology course students over the last five years,
who originally necessitated, and in a sense
fostered, this book. I am also grateful to
numerous referees of my manuscript, particularly
Robert Chapman of Johns Hopkins University.

I was regrettably unable to follow many points of
advice offered by these referees, so that the final
result is my responsibility. Parvaneh Rafiee
prepared the figures, while Mary Primrose
photographed them. My greatest debts are to Michel
Krieber and Diane Edmonds, who went over the
manuscript with great care, finding numerous errors
that no one else, myself included, had detected.
The two of them have also been the best possible
source of inspiration, readers who are at least
willing to act as if a book has been of enduring
value to them.

Department of Biology,
Dalhousie University,
Halifax, Nova Scotia

INTRODUCTION

This is a "how-to" book. Its purpose is to teach students with some general background in ecology how to use the elementary mathematical models which underlie many of the important scientific themes of ecology. In order to realize this goal, it is vital that the student develop the capacity to think in terms of mathematical models, rather than simply taking the implications of a model "on faith". For undergraduate students of physics or chemistry, the development of such understanding is an indispensable part of their education. The tradition in biology has been different, however. Relatively few biology lecturers have much mathematical background themselves, and in any case most sub-disciplines of biology do not use mathematical tools in developing their theories. The major exception to this pattern is population biology. Whether the concern has been evolution or ecology, population biologists have for some time approached the theoretical issues using quantitative mathematical models. Advanced undergraduate or graduate students interested in evolution or ecology have had little choice but to acquire some kind of familiarity with the use of mathematical models.

All too often, that familiarity has been developed on an ad hoc basis, the problem being the lack of systematic teaching of mathematical modelling in population biology. Indeed, many of those biology students who do acquire some understanding of mathematical models obtain it from courses in theoretical physics or applied mathematics. In population genetics, there are a number of excellent books which teach the basic models in a thoroughgoing fashion. My personal favourite is Nagylaki's (1977) Selection in One-

and Two-Locus Systems. Books like this are self-contained packages which can be used to develop the mathematical facility required for dealing with the extant theoretical work on evolution. The present book is an attempt to do the same for those who want to be able to deal with ecological theories in a quantitative, mathematical fashion. As such, it is in no sense a "textbook of ecology", or even properly an introduction to theoretical ecology. There are a number of books which fulfill those functions admirably. Such books cover "who-did" and "in-fact" functions, telling the student about the history and the empirical successes of ecology. So long as such material is to the fore, it will be difficult to convey the way of thought required to take on the theory on its own terms. Experience suggests that it is in fact impossible to do so. Traditionally, physics departments have given their majors separate courses in the mathematics of classical mechanics, not primarily to teach them the history of the field, but to cultivate the ability to develop and analyse mathematical models. The present book is intended to achieve precisely this end for biology students with interests in ecology.

A word or two might be said about the demands which will be made of the reader. Firstly, it is hoped that it will not be necessary to have several reference works on hand while using this book. Considerable effort has been made to keep it self-contained. Secondly, it is not intended to require additional lectures. It is hoped that the book could function as an auxiliary to conventional ecology lectures and texts, as a tool of independent study. Thirdly, the book cannot simply be read. It must be worked through. This means more than the solution of the problems at the end of each chapter. Any passage in the text which the reader cannot understand should be approached as a puzzle to be solved. The solutions required for understanding will lead to advances in the reader's ability to deal with mathematical models in the ecological context, and thus are critical to the purpose of the book. This means that the reader must have some patience, some willingness to struggle to understand. Fourthly, it is inevitable that there are some prerequisites where mathematics are concerned. It is assumed that the reader has had enough exposure to calculus to find the operations of differentiation and integration somewhat familiar, even if the reader has forgotten

most of what was once learned. (I endeavour to
review relevant material briefly as I proceed.)
Some acquaintance with analytical geometry and
matrices would also be helpful, although the
present treatment presumes little working ability,
at least at the outset. However, it should be
emphasized that this is not a mathematics textbook.
My primary aim is to teach modelling, which
necessarily involves some mathematics, but
primarily involves the development of skills which
are not normally taught in mathematics departments,
skills which are used chiefly by scientists and
engineers.

THEORETICAL MODELS IN ECOLOGY

In fields like physics, the role of mathematical
models is well-defined. A collection of hypotheses
is gathered together, normally taken from the
standard, widely agreed upon, hypotheses available
at the time. Inverse-square laws of gravitation
would be an example of such a hypothesis. A
particular problem is then chosen, such as the
orbit of one body around another. A model is
then developed to represent that situation, with
general variables in place of specific parameter
values, and analysed with a view to unravelling the
ways in which the "behaviour" of the model depends
on the parameters of the model. To test the model,
a specific instance of such a situation is chosen,
an example in this instance being the orbit of the
Moon about Earth. The values of the model
parameters are estimated by means of observations
on this particular system, and these values are
plugged into the model in order to generate
specific predictions concerning the shape, the
periodicity or whatever, of the orbit. The actual
orbit is then observed to see if the model's
predictions are met. If not, then it is assumed
that, if no technical error has arisen, one of the
initial hypotheses was incorrect. The hope is that
the incorrect hypothesis can be identified and a
correct replacement found.
　　Readers with some background in ecology will
realize that ecology does not proceed in anything
like this fashion, with the direct testing of
quantitative predictions of explicit models
deciding critical issues. There are two obvious
questions. Firstly, why isn't this method used in

ecology? Secondly, given that it isn't, what
possible value do mathematical models have in
ecology? These questions will be dealt with in
turn. Firstly, physics treats processes which
normally possess very consistent properties: laws
such as conservation of mass-energy and universal
fixed constants such as the speed of light in a
vacuum. Thus physicists are able to given single
universal numbers to many of their model
parameters, and use a fair-sized collection of
well-corroborated hypotheses, often called "laws",
in formulating their models. Even a modest
acquaintance with ecology reveals that it is not
possible to assume universality of parameter
values, nor are there any indisputable "laws".
There is a general lack of "structure" and a
surfeit of variability between particular
ecological systems. Thus the prospects for drawing
firm conclusions about the validity of specific
hypotheses from the predictive success of
particular models which embody those hypotheses are
slim indeed.

Secondly, given this situation, isn't it fair
to conclude then that mathematical models have no
value in ecology? Wouldn't we be better off just
using our intuition? The problem with this line of
thought is that science can not normally get very
far in understanding quantitative dynamics without
mathematics. Certain branches of biology, most
notably molecular biology, do quite well on the
basis of syllogisms alone. But the questions that
specialists in these areas tackle are of an
all-or-none kind, or involve counting a finite
number of things. What is the molecule responsible
for heredity? How many copies of a gene are there?
If ecology is indeed about the distribution and
abundance of organisms (cf. Krebs, 1985), then it
is difficult to see how quantitative reasoning can
be avoided. The alternative is not to reason at
all. Such quantitative reasoning in ecology cannot
hope to achieve the degree of quantitative
stringency characteristic of physics, and the
attempt to do so must be seen as a cardinal error.
The extent to which mathematical models in ecology
can "stretch" the limits of our understanding will
be illustrated throughout the present introduction
to the basic models of theoretical ecology. This
is not to say that any of the models to be
discussed here have much concrete applicability.
[Ecological models having such applicability are
usually much more complex than those discussed

here, often being embodied only in computer simulations.] But it should become obvious to the reader that many of the possibilities which ecological models reveal are difficult to take into account at the level of verbal reasoning alone.

MODELS COVERED HERE

No book designed to teach mathematical modelling in ecology to those with no prior background in modelling can cover all the theory which has been developed in ecology. My approach has been to confine myself to deterministic models which are readily discussed in terms of dynamical systems theory (Hirsch and Smale, 1974) without invoking stochastic, or "probabilistic", processes. Models of this kind were the first proposed in theoretical ecology, chiefly by Lotka and Volterra, and have since played a central role in the development of theoretical ecology (e.g. May, 1974). They are also the most elementary mathematically, other types of model often using them as constituent components.

Within the general context of deterministic models in ecology, I have tried to cover the basic models fairly generally. Both "continuous-time" and "discrete-time" models are discussed, since it cannot be generally assumed that the former will always be sufficient. [Don't worry if these terms have little meaning now; they will be explained soon.] Models of ecosystems with one, two, three, or many species are considered, although as the number of species increases, the scope of analysis steadily falls, for reasons which will be made clear. The conventional ecological processes are considered: population growth, competition, predation, mutualism, and migration. Of the major theoretical "issues" in ecology, the problems of ecosystem stability, coexistence of competitors, chaos, predator-prey cycles, and multiple stable-states are all considered, because they arise naturally from any discussion of the models covered here.

However, it should be repeated that this book is not a general text. Perhaps the closest analogue would be a laboratory manual, technically oriented and lacking a discursive overview of the field. But just as it is difficult to be a zoologist without acquiring some technical knowledge of anatomy in a laboratory dissection course, so it is difficult to be a well-rounded ecologist without some ability to handle ecological models with understanding, rather than mystified diffidence.

Chapter One

POPULATION GROWTH

We begin with the most elementary idea in ecology:
the growth in the number of individual organisms
making up a single population. This is not to say
that it is assumed that this population does not
interact with other populations. Such interactions
will be allowed <u>implicitly</u> in the way the models
will be built, without explicit representation of
other populations, as will be discussed.

1.1 LINEAR CONTINUOUS-TIME MODELS

The models that most people are exposed to first
when they are taught ecology are "continuous-time"
models of the growth of single populations. Models
of this kind were also the first to be developed in
ecology, having origins in the eighteenth and
nineteenth centuries in the works of Malthus and
Verhulst. In fact, such models were among the
first formulated in the development of physics too,
leading Newton to develop calculus, the fundamental
mathematics necessary to handling such models.
        The central assumption of continuous-time
models is that time passes as an unbroken stream,
and is thus "continuous", rather than passing in
discrete quanta, or stages. This is of course the
way most Western people think of time, when they
think about it at all. Another implicit assumption
in such models is that time can proceed
indefinitely, without end. Again, this is an
"everyday reality" assumption made by most people.
Like time, it is also normally assumed that all the
properties of the thing being modelled can be
represented by continuous variables. Your mass,
for example, would be considered as a continuous
variable in most physics models, as would the
distance from your eyes to this page.

In ecological models of population growth,
population density is taken to be a continuous
variable of the same kind. Strictly speaking, this
assumption is wrong. Let us consider a specific
terrestrial habitat of area A, with a total of N
animals of the species <u>Spurious</u> <u>fictitious</u> present
in the habitat. While A is certainly a continuous
variable, N is not. There can't be 786.12 animals
in the habitat, there can only be 785, 786, 787,
and so on. Population density similarly cannot be
a continuous variable. Say A is 100 square miles,
then the population density will vary in units of
.01 animals per square mile. Population density
cannot take on values like 7.8612 animals per
square mile for the same reason that there cannot
be 786.12 animals in the habitat. However,
allowing the existence of .12 animals allows the
use of simpler mathematics than would otherwise be
possible, and one of the central factors in
developing mathematical models in any science is
the need for simplicity. If it doesn't have
simplicity, it is normally impossible to use a
model at all. Thus ecologists almost always take
population density to be a continuous variable.
    Let us define a population density variable,
then. We will take

$x = N/A$ (1.1)

where A is the area or volume of the habitat of
interest and N is the total number of organisms of
the species of interest, both allowed to vary
continuously, like population density.

The "Malthusian" or Density-Independent Model

If we are interested in population growth, then we
must be allowing for the possibility that x changes
with time. Thus, in mathematical terms, x is said
to be a function of time, which is represented
explicitly by x(t). [Similarly, we must allow that
N is a function of time, N(t), too.] By contrast,
it would be normal to assume that A is fixed, or so
slowly changing that its "dynamics" [patterns of
change] can be neglected. However, once we have
agreed to make these variables functions of time,
it is often convenient to suppress "(t)" when we
write them down, if only for the sake of saving
space.
    In almost all cases, when we develop a
mathematical model in ecology, we will want to know
how the pattern of a variable's change with time,
its dynamics, depends on specific assumptions about
the ecological processes which affect it. So we

need some formal way of proceeding from such assumptions to conclusions about the dynamics of the model variables.

Let me offer an analogy which might help you think about this problem in concrete terms. Questions about dynamics are rather like questions about the route travelled on a journey. Say that the journey is being taken by a robot-driven car. We could specify that the car proceeds back and forth along a completely straight highway. Then, if we assume that zero represents the starting point on the highway, the distance travelled in a given time will depend on the speed that the car is driven at, where this speed may vary. Measurements of speed refer to the distance travelled per unit time, such as so many miles per hour. But we don't have to wait an hour and check the odometer to know what the speed of a vehicle is. We can look at the speedometer instead. This gives the instantaneous rate of change in position, the speed. In terms of calculus, if y represents the position of the car along the highway and t represents time, the speed is dy/dt , the "d's" referring to conventional differentiation of y with respect to t .

Similarly, a population has a rate of change in its density, the population "growth rate", where this growth rate is fully analogous to the speed of a car. Mathematically, this is represented as dx/dt , the change in the population density per unit time. If we can find an expression for this quantity, then we may be able to determine how population density changes with time.

Imagine a population growing in an environment without limits, with no particularly important interactions between individuals within this environment. [Bacterial ecology might sometimes be this simple.] Then we might assume that the only determinants of population growth are a fixed birth rate and a fixed death rate. [Most importantly, such birth and death rates will be taken as "density-independent", in that an individual organism's survival probabilities and fecundity are assumed to be unaffected by the number of other organisms of that species in the immediate habitat.] Say that births and deaths occur at rates b and d per individual, respectively. It is now reasonable to write the following equation:

$$dN/dt = bN - dN = (b-d)N \qquad (1.2)$$

Evidently, the assumption of fixed rates of birth and death per individual makes the growth of the population dependent on the size of the population.

Perhaps equation (1.2) will be more acceptable if we define

$$r = b - d \qquad (1.3)$$

giving

$$dN/dt = rN \qquad (1.4)$$

an equation which should be familiar from introductory ecology. The parameter r is known as the "Malthusian parameter" or the "intrinsic rate of increase". Recall that $x = N/A$, equation (1.1), and our assumption that $dA/dt = 0$. A basic law of calculus is that, if we have a variable which is equal to a ratio of two other variables, say $e = f/g$, then

$$de/dt = [g \times df/dt - f \times dg/dt]/g^2$$

Taking all this into account, we can derive dx/dt as follows:

$$dx/dt = d(N/A)/dt = (A \times dN/dt - N \times dA/dt)/A^2$$

$$= (dN/dt)/A = rN/A = rx \qquad (1.5)$$

Equation (1.5) is apparently an exact parallel to equation (1.4), so that population growth in models like this can be treated in the same way whether density or numbers are used.

In case the context hasn't made it clear, an equation like (1.5) is regarded as a model of population growth. This particular model embodies the assumption that birth and death rates are fixed. Given this assumption, and thus the model, what is the next step? The ideal solution to a mathematical model of temporal dynamics is to come up with a function which explicitly predicts the values that the model variables will take on at any particular time. [As a practical matter, this requires that we eliminate the dx/dt expression.] In the present case, we would like to find some $f(t)$ such that $x(t) = f(t)$. This can in fact be done. We have

$$dx/dt = rx ,$$

which can be rearranged as

$$(1/x)dx = r \, dt$$

The next step is to integrate both sides, as follows:

$$\int (1/x) \, dx = \int r \, dt \qquad (1.6)$$

Looking in a table of integrals reveals that integration of $1/x$ with respect to x gives $\ln x$, the natural logarithm of x -- meaning that the base is e instead of 10, while integration of r with respect to t gives rt. Thus equation (1.6) can be written as

$$\ln x + c_1 = rt + c_2 \qquad (1.7)$$

10

where the c_i are arbitrary constants which can be used to fit a "solution" trajectory to appropriate initial conditions.

A word might be said here about integration for those who haven't seen any integral calculus for a while. Integration is really a way of summing up the values taken on by a function. To return to our robot-driven car, as it travels along the road at some speed, it is accumulating distance with respect to its starting point. If its speed at any given time is represented as s(t), then

$$\int_0^\tau s(t)\ dt$$

gives the total distance travelled between t = 0 , the starting time, and t = τ , the time the journey ended.

Thus it is natural to use integration to calculate how much the population will have grown after a given amount of time has elapsed. Equation (1.7) can be simplified by defining c = c_2 - c_1 , giving

$$\ln x = rt + c \qquad (1.8)$$

Since ln is the base e logarithm, if we take the exponential function of both sides, we can get an expression which just has x on the right-hand side, or "RHS" as we will abbreviate it. Thus we obtain

$$x = \exp[rt+c] = De^{rt} \qquad (1.9)$$

where D = exp[c]. We would like to have an interpretation of D, since in equations which are used in scientific modelling we cannot have meaningless symbols cluttering up the page. Consider equation (1.9) when t = 0 . Then x(t) = De^0, but e^0 = 1 , so x(t) = D when t = 0 . Thus the solution to the model embodied in equation (1.5) is

$$x(t) = x(0)e^{rt} \qquad (1.10)$$

This gives us a formula explicitly representing the dynamics of x with respect to time. If we know x(0) and r , we can calculate x(t) for any value of t .

If ecology were a field like physics, we could stop here and wait for further information, specifically estimates of these parameters. But it is practically very difficult to estimate parameters like x(0) and r in real populations, even in those special cases where models this

simple have some validity. So we use a different
approach in interpreting our models in ecology. We
ask how the predictions of our models change
qualitatively with substantial changes in the model
parameters, because we have some hope of detecting
major differences in model parameters or actual
ecosystem behaviour. Small quantitative changes in
either are unlikely to be detected.

In the case of the predictions obtainable from
equation (1.10), we want to know what sort of
parameter changes make major differences. Let us
consider variations in the value of r . If we
have $r > 0$ then as t approaches ∞ , represented
as $t \to \infty$, the value of e^{rt} approaches ∞ too, or
$e^{rt} \to \infty$, because as y becomes arbitrarily large
the limiting value of e^y is ∞ . If $r = 0$, then
$e^{rt} = 1$ always, since $e^{r0} = e^0 = 1$. If $r < 0$,
then as t approaches ∞ , or $t \to \infty$, the value of
$e^{rt} \to 0$, because as y becomes arbitrarily large
$e^{-y} = 0$. [More exactly, $\lim_{y \to \infty} e^{-y} = 0$.]

Given this pattern of change in the e^{rt} function,
there are only two values of $x(0)$ that matter:
$x(0) = 0$ and $x(0) > 0$. [These are called
"initial values".] Irrespective of r , if $x(0) = 0$
then $x(t) = x(0) = 0$ for all values of t .   If we
have $x(0) > 0$, then there are only three eventual
outcomes, which depend on the value of r . If $r >$
0 , then $x(t)$ will increase without limit. If
$r = 0$, then $x(t) = x(0)$ for all values of t .
If $r < 0$, then $x(t)$ will decrease toward 0 as
time proceeds; that is, population growth will be
negative.

Practically speaking, what does all this mean?
Firstly, nobody is going to be studying a
population which has an initial population density
of zero, since the population wouldn't be there to
study. Therefore we can neglect the possibility of
$x(0) = 0$. Secondly, while such cases sometimes
need attention in the mathematical context, no real
population is going to have a value of r precisely
equal to zero. [In fact, r is unlikely to have
any single precise value for an indefinite period
of time, but that is a problem that we will come to

later.] Thus there are only two cases to be
considered, either r < 0 or r > 0 . In the
first case, the population density falls toward
extinction. In the second case, the population
density increases explosively. These two cases are
displayed graphically in Figure 1.1. In fact,
numerical calculations using an equation of the
same type as equation (1.10) with r > 0   were the
inspiration for Thomas Malthus's (1798 [1926])
Essay on the Principles of Population, in which he
predicted ever-increasing human misery unless we
refrained from sex. This is the reason why model
(1.5) is referred to as "Malthusian".

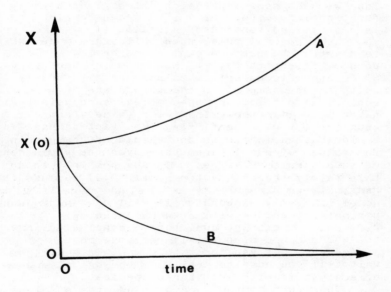

Figure 1.1. The pattern of
population growth when r > 0   in
model (1.5) is illustrated by
curve A. The pattern of population
growth when r < 0 is illustrated by
curve B.

The Logistic Model

Now with the slightest knowledge of ecological
thinking on questions of population growth, you
will regard model (1.5) as pathetically inadequate
as a description of the way real populations grow.
Your mind is probably thinking about the exhaustion
of food and space as population growth proceeds

toward population densities of 10^{12} and beyond.
You may even have alternative equations for dx/dt
in mind. Most of these will share with your
intuition the supposition that r should really be
a function of the population density of the
species, so that, as population density increases
to sufficiently high levels, the death rate rises
and the birth rate falls, bringing about a fall in
r toward zero and below, into negative values.
Thus we need some expression which gives us a
functional form for $r(x)$, in order to plug it
into a revised version of equation (1.5).

As a general rule, when building a model which
is meant to be used to explore general
possibilities, it pays to make simple assumptions.
Since specific numerical values are not usually
going to be assigned to the parameters of
ecological models, any theoretical conclusions will
have to be like those that we obtained from
equation (1.10), identifying important changes in
ecological variables as the result of changes in
the values of model parameters over the whole range
of such parameter values. It was possible to do
this fairly thoroughly for model (1.5) because it
had just two parameters, $x(0)$ and r . If a
model is overly elaborate, it will have many such
parameters, and it will be correspondingly
difficult to extract qualitative conclusions over
the whole range of possibly relevant model
parameters. [If you don't believe this now, just
wait till you have to deal with ecological models
of ecosystems with just three species.]

To return to the case at hand, the simplest
function for r(x) other than the constant value
assumed in model (1.5) would be a linear function,
having the general form

$$r(x) = a - bx \qquad (1.11)$$

In particular, assuming that the population grows
healthily at low densities, we could take $a > 0$.
On the other hand, it seems reasonable to assume
that $b > 0$, reflecting the kind of
"self-limiting" density-dependence which is called
to mind whenever the "Malthusian nightmare" is

summoned up in our imaginations. Now we substitute
the expression for r from equation (1.11) into
equation (1.5), giving

$$dx/dt = (a-bx)x = ax - bx^2 \qquad (1.12)$$

The problem with model (1.12) is that the
parameters a and b are fairly meaningless. To
find a meaningful interpretation of them we will
consider two special cases of model (1.12).

(a) First, consider the case of low densities,
when $x \approx 0$. In this case, we have

$$dx/dt = ax - bx^2 \approx ax$$

But r(0) = a , so

$$dx/dt|_{x \approx 0} \approx r(0)x \qquad (1.13)$$

[The $f(x)|_{x=z}$ notation means the value of f when
x = z . Sometimes we will also represent this as
f(z) .] This gives us our interpretation of a ; it
represents the rate of population growth when
density-dependent limitation is absent.

(b) Second, consider the case when dx/dt = 0
but we have x > 0 . For this case to arise, we

need $0 = ax - bx^2 = r(0)x - bx^2$, for some
particular x , call it x* . If x* is not equal
to zero, then this requires r(0) - bx* = 0.
[This comes from factoring an x out of the dx/dt
equation.] Rearranging this expression, we must
have x* = r(0)/b . From your other ecology books,
you will have heard the term "carrying capacity",
which refers to the population density of a
particular species that can be sustained in its
ecosystem. You should check that for x > x* ,
dx/dt < 0 , while for x < x* , dx/dt > 0 .
Therefore, x* is the carrying capacity, normally
represented as K . Rearranging our equation for
x* , we have

$$b = r(0)/K \qquad (1.14)$$

Substituting equation (1.14) back into equation
(1.12), we get

$$
\begin{aligned}
dx/dt &= r(0) - [r(0)/K]x^2 \\
&= r(0)x[1 - x/K] \\
&= r(0)x[K-x]/K \\
&= rx[K-x]/K \qquad (1.15)
\end{aligned}
$$

where this last form seems more reasonable, since
we have disposed of the need for the expression
r(x) as such. Equation (1.15) is in fact the
famous "logistic" equation for population growth,
so often discussed in ecology texts. Whatever its
empirical merits, it is the mathematically simplest

representation of density-dependent population regulation. This equation has several nice features from an intuitive standpoint. Firstly, when x is small, the fraction $[K-x]/K$ is about one in value, giving essentially Malthusian population growth. This pops right up from the way the equation is written. Secondly, if r is positive, then the sign of dx/dt depends on the value of x relative to K. For $x > K$, $K - x$ is negative, while for $x < K$, $K - x$ is positive, suggesting that the population growth trajectory should "home in on" the population density given by K. [You will need to get used to developing this sort of interpretation of model parameters as we go along.]

Having set-up our model and found an interpretation of it, we now proceed to solve it. Following the same kind of approach as the one we used for the Malthusian growth model, we split up dx/dt , giving

$$Kdx/x[K-x] = rdt \qquad (1.16)$$

We want to integrate both sides again. We already know the integral for the RHS (right-hand side) of the equation, but the LHS is pretty messy. One way to solve this problem is to split the LHS up into a sum of fractions. We would like to find constants D and E such that

$$K/x[K-x] = D/x + E/[K-x]$$

For such D and E , we must have $D[K-x] + Ex = K$. So we must have $D = 1$, since it is the only one of the two constants which is multiplied by K . With $D = 1$, it is obvious that E must be 1 too. Thus equation (1.16) becomes

$$\{1/x + 1/[K-x]\} dx = r dt$$

Now we integrate both sides, as before

$$\int 1/x \, dx + \int 1/[K-x] \, dx = \int r \, dt$$

finding

$$\ln x - \ln K-x = rt + F$$

which can be rearranged as

$$\ln x/[K-x] = rt + F$$

where F is an arbitrary constant. Again, we take the exponential function of both sides, finding

$$x/[K-x] = Ge^{rt} \qquad (1.17)$$

where G is also an arbitrary constant. We should again find out what G is. If we take $t = 0$, the e^{rt} term will disappear from the equation, giving

$$G = x(0)/[K-x(0)] \qquad (1.18)$$

Substituting equation (1.18) into (1.17) will give

$$x(t) = Kx(0)e^{rt}/[K - x(0) + x(0)e^{rt}] \quad (1.19)$$

as you are to show in one of the Exercises.

Having found a solution for our model, we can go on to investigate its predictions concerning the course of population growth. The question which is of most interest in this context is the eventual fate of the population. Mathematically, this can be found by seeking to discover what happens to $x(t)$ as $t \to \infty$, which means finding the "limit" of $x(t)$ as t approaches ∞. In the case at hand, we seek

$$\lim_{t \to \infty} x(t) = \lim_{t \to \infty} \{Kx(0)e^{rt}$$

$$\qquad \div [K - x(0) + x(0)e^{rt}]\}$$

$$= \lim_{t \to \infty} \{K/[e^{-rt}/G + 1]\}$$

$$= K/\{\lim_{t \to \infty} [e^{-rt}]/G + 1\}$$

$$= K/\{0/G + 1\}$$

$$= K \quad (1.20)$$

This shows that, providing $x(0)$ is greater than zero, we have the population growth process taking the population density to the value K, just as our intuitive analysis of the logistic equation suggested. The types of population dynamics that the logistic equation can give rise to are shown in Figure 1.2. This sort of behaviour is called "convergence to an equilibrium". The point $x = K$ is considered an equilibrium because there $dx/dt = 0$. It is converged to because $\lim_{t \to \infty} x(t) = K$,

which is the meaning of the word convergence in the modelling context. Equilibria that are converged to are called "asymptotically stable" equilibria. The word stable is used because once the population density is near K, it is not likely to move very far from it. K is a kind of sticking point for the population growth trajectory. Another word for it is "attractor", since it seems to attract the trajectory to move toward itself. This notion of stability will feature prominently in our discussion, so you will have many more chances to develop your understanding of it.

It may have already occurred to you that the logistic model assumes that population growth proceeds quite nicely when there are only 10^{-11} organisms per unit of habitat. In animal species with two sexes, at least, one has to wonder how so few animals, possibly less than two, can manage to get together to mate. Another problem is that this model assumes that an increase of one more intraspecific competitor is equally bad, whether there are 100 animals in the habitat or 10 trillion. But surely things are not so uniform, surely things are not so linear? Indeed, one of our most interesting tasks will be to investigate the effect of relaxing assumptions of this kind, to make our models more realistic, which is good, but also more elaborate, which is bad. The technical problem is how to continue figuring out the implications of our models even when they are fairly elaborate.

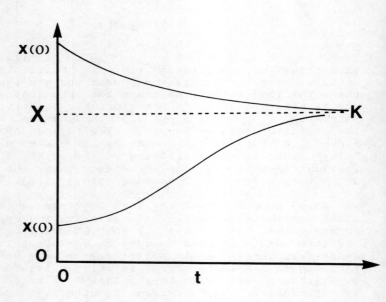

Figure 1.2. Convergence to the carrying capacity, K , which is exhibited by all trajectories of x(t) , whatever the value of x(0), providing x(0) > 0 .

1.2 NONLINEAR CONTINUOUS-TIME MODELS

All the models that we have discussed so far have
had simple functions relating population growth
rate to population density. Mathematically, such
models are extremely attractive because they are
easy to analyse. Empirically, such models are
attractive because it is much easier to estimate
constants or perform linear regressions than it is
to estimate the parameters defining nonlinear
equations. But you are not likely to suppose that,
just because it is convenient to do so, ecological
dynamics must always depend on linear functions, in
general. On the other hand, if we suppose that
ecological interactions are generally nonlinear,
then we must admit that it is unlikely that we will
ever have much precise knowledge about their
dynamics in particular cases, given the empirical
difficulties. Where does this leave the theory?
As it turns out, we often do not need to know the
precise expressions defining a population's
dynamics, providing we have certain basic types of
qualitative information. This is because the
mathematics give fairly simple results, at least
where the long-run dynamics are concerned, as will
now be shown.

General Autonomous Models

In considering population growth models in
which dx/dt is not a linear function of population
density, there are two broad categories of model
which need to be distinguished. The first of these
includes all models which specify dx/dt with
reference to parameters which do not themselves
depend on time. Such models are called
"autonomous". The second category, naturally
enough, is made up of those models for which this
is not true, time being one, if not the only one,
of the parameters which determines population
growth rate. Such models are called
"nonautonomous". We will discuss each type of
model in turn.

For single-species population growth, the
general continuous-time autonomous model has the
following form:

$$dx/dt = r(x)x = f(x) \qquad (1.21)$$

Unfortunately, it is often not possible to find a
solution of this model for the trajectory of x(t)
having the form

$$x(t) = F(t) \qquad (1.22)$$

None the less, it may still be possible to develop

a fairly complete understanding of x(t) without a solution of the form of equation (1.22), providing we know enough about the RHS of model (1.21). Most importantly, if we can always differentiate r(x) with respect to x , then we can at least characterize the asymptotic behaviour of x(t) qualitatively.

Consider a particular example of a nonlinear r(x) function, that shown in Figure 1.3. There are

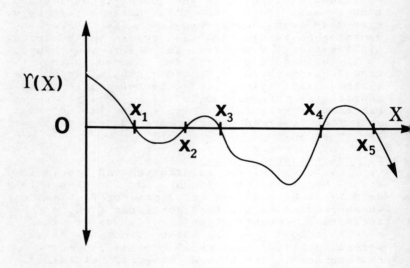

Figure 1.3. A particular example of a nonlinear r(x) function for model (1.21). For this function, it is assumed that r(x) < 0 for all x > x_5 .

methods for analyzing the dynamics engendered by

functions of this form. One is to find the "Taylor expansion" of the function $f(x)$ from (1.21) near equilibria, when $f(x) = 0$. Say that there is an x^* such that $f(x^*) = 0$. Then the Taylor expansion of $f(x)$ in the neighbourhood of x^* is given by the following equation:

$$f(x) = f(x^*) + (x-x^*)df(x^*)/dx$$
$$+ (x-x^*)^2/2! \ d^2f(x^*)/dx^2 + .$$
$$. + [(x-x^*)^n/n!]d^nf(x^*)/dx^n$$

(1.23)

The notation for the factorials and derivatives should be familiar, but for those who need some reminder $n! = n(n-1)(n-2) . . 1$, so that $4! = 24$

for example, while $d^nf(x^*)/dx^n$ refers to the value of the formula obtained by differentiating $f(x)$ n times and evaluating the result at $x = x^*$, so that

$d^2[x^3]/dx^2$ at $x = 5$ is 30 . Note that the smaller the distance between x and x^* , the smaller the value of $x-x^*$. For $|x-x^*| << 1$, we can write

$$f(x) \approx (x-x^*)df(x^*)/dx$$

(1.24)

to a good approximation, because the $(x-x^*)^n/n!$ factors become quite small for large values of n. [If your quantitative imagination boggles at this, calculate 0.1^5.] The sort of situation when this approximation will be in considerable error is when $df(x^*)/dx$ is smaller than

$d^2f(x^*)/dx^2$. When this is the case, more terms in the Taylor expansion need to be used in any approximation of $f(x)$. [These are called "higher-order" terms, because the value of n is larger.] The value of the estimate given by equation (1.24) is that it allows linearization of models like (1.21) near equilibrium values.

This is done as follows. Given that (1.24) is indeed an adequate approximation, we set up a differential equation for

$$\epsilon = x - x^*$$

(1.25)

where from this definition

$$d\epsilon/dt = dx/dt - dx^*/dt = dx/dt$$

(1.26)

Near x^* , where $x \approx x^*$, ϵ is obviously small and we can take

$$d\epsilon/dt = dx/dt = f(x) \approx (x-x^*)df(x^*)/dx$$

giving

$$d\epsilon/dt = \epsilon \times df(x^*)/dx \qquad (1.27)$$

If $df(x^*)/dx < 0$, then $d\epsilon/dt < 0$ for $\epsilon > 0$ and $d\epsilon/dt > 0$ for $\epsilon < 0$. Thus $|\epsilon(t_1)| >$

$|\epsilon(t_2)|$ for $t_2 > t_1$, and we have $\epsilon(t)$

approaching 0 as t approaches infinity, giving x(t) approaching x* as t approaches infinity. In this case, then, x* must be an asymptotically stable equilibrium, or attractor, at least in the immediate neighbourhood around it. In the opposite case, if $df(x^*)/dx > 0$ then $d\epsilon/dt > 0$ for $\epsilon > 0$ and $·d\epsilon/dt < 0$ for $\epsilon < 0$. From this, using the same sort of argument as before, we will have $|\epsilon|$ increasing with time near x* . It would be incorrect to conclude from equation (1.27) that $\epsilon(t)$ approaches infinity as time approaches infinity, because that equation is only valid so long as approximation (1.24) is valid, which will not necessarily be true for large ϵ . However, it is obvious that in this case the equilibrium cannot be an attractor, that is asymptotically stable.

In any event, "locally" (meaning near x*) the dynamics hinge on the sign of $df(x^*)/dx$, and thus the slope of f(x) near values of x for which f(x) = 0 . With this in mind, consider Figure 1.3. We can see that there are the five equilibria indicated by x_1 to x_5 , in addition to the

equilibrium with x = 0 , because r(x) = 0 at these points. It should be borne in mind that
$$df(x)/dx = d[r(x)x]/dx = x dr(x)/dx + r(x)$$
When x = 0 , df/dx is positive, in the case given by Figure 1.3, because r(0) is positive. In all the other cases, r(x) = 0 and x is strictly positive, so that the sign of df/dx is given by the sign of dr/dx . Therefore the dynamics about all these equilibria can be readily inferred simply from whether r(x) is increasing or decreasing as it crosses the x-axis. Thus we can see directly from Figure 1.3 that 0 , x_2 ,

and x_4 all define unstable equilibria. On the

other hand, x_1 , x_3 , and x_5 are all locally

stable equilibria. The obvious question, then, is what are the population growth dynamics like, in general, for all initial values of population density? For x(0) in the interval $(0, x_2)$, x(t)

22

converges to x_1 ; for $x(0)$ in (x_2, x_4) , $x(t)$
converges to x_3 ; and for $x(0)$ in (x_4, ∞) , $x(t)$
converges to x_5 , as is to be shown in the
Exercises. In effect, the unstable equilibria act
as thresholds demarcating the regions within which
the population density converges to different
attractors. In any case, given graphs of $r(x)$
functions like that of Figure 1.3, we can discover
most of the essential properties of an autonomous
model's asymptotic behaviour.

Density-Independent Nonautonomous Models
If changes in the population growth rate are
imposed by the environment in a fashion which is
both time-dependent and density-dependent, it is
difficult to say much about the analysis of
appropriate models with much incisiveness. However,
when population growth rates are only
time-dependent, that is nonautonomous, without
density-dependence, then it is possible to discuss
the analysis of such models profitably and remain
at a reasonably simple level. The appropriate
models will have the following form:
$$dx/dt = r(t)x \qquad (1.28)$$
This model can be solved in the same way that the
Malthusian and logistic models were solved.
Rearranging equation (1.28), we find
$$1/x \ dx = r(t) \ dt$$
As before, we can integrate both sides

$$\int 1/x \ dx = \int r(t) \ dt$$
giving
$$\ln x = [\int r(t) \ dt \ / \ t]t + C$$
Now in fact the expression $\int r(t) \ dt \ / \ t$ is the
average value of r over all times. If we let
$$\rho = \int r(t) \ dt \ / \ t \qquad (1.29)$$
then
$$x(t) = x(0)e^{\rho t} \qquad (1.30)$$
This result is obviously analogous to that found
for the Malthusian model, as represented by
equation (1.10).

This model gives rise to population dynamics
like those of (1.10) for $\rho \ll 0$ or $\rho \gg 0$, low
or high asymptotic densities, respectively. But
for $\rho \approx 0$, it is possible that there could be
empirically important fluctuations of population
density. There is one point which deserves special
comment in this case. Since this is a

continuous-time model, for $x(0) > 0$, any transient
negative values of $r(t)$ will not drive the
population to extinction, because that occurs only
in the limit, as time approaches infinity, even
when $\rho \ll 0$. But since we have $\rho \approx 0$, if $r(t)$
is much less than zero for some particular $t = \tau$,
then there must be some value Δ such that $r(\tau+\Delta)$
> 0 , so that we cannot have $x(t)$ asymptotically
decreasing toward zero for $\rho \approx 0$. A similar
argument can be used to show that $x(t)$ cannot
increase without bound. Thus the population
fluctuates over the open interval $(0,\infty)$. In
effect, there are three distinguishable alternative
predictions from this model, or at least three
alternatives which are distinguishable in principle
given that there is enough time for observation.

1.3 DISCRETE-TIME MODELS

Our second group of models will be based on the
assumption that single-species population growth
occurs in discrete quanta of time. This leads to
models based on finite-sum calculus, particularly
<u>difference equations</u>, the discrete-time analogues
of differential equations. Your intuition may
wonder how it could be appropriate to consider time
passing in discrete quanta, at least from an
ecological standpoint. One context where this
might make more sense to you is to consider the
population growth of small temperature-sensitive
organisms in temperate habitats, like those of
southern Canada, in which winter must be passed in
the form of a resting stage which carries out
little metabolic activity, and certainly does not
reproduce. Life is essentially lived only in the
summer, when for a few weeks or months the life
cycle must proceed from egg to larva to adult and
on to egg again, in the case of many univoltine
insects, or similarly from seed to plant to seed,
in the case of many annual plants. For such
organisms, it may be most appropriate to consider
the population growth process as one occurring in
discrete steps, each step corresponding to one
year. Again, we will specify our variable of
interest as the population density of a species in
a delimited habitat, represented by X , and
proceed to run through the immediately obvious
simple models.

<u>Density-Independent Model</u>
As mentioned, the basic formalism used to model

population changes which are supposed to occur in
discrete quanta of time is the difference equation,
which in the case of single-species population
growth has the following form:

$$X(t+1) = g[X(t),t]X(t) \qquad (1.31)$$

The simplest possible form for g is a constant,
say R , giving

$$X(t+1) = RX(t) \qquad (1.32)$$

where we must have $R \geq 0$ for realism, since
negative values of X would make no biological
sense. This model is almost indecently easy to
solve. If equation (1.32) gives us X(t+1) , then

$$X(t+2) = RX(t+1) = R[RX(t)] = R^2X(t)$$

giving, by inductive generalization,

$$X(t+n) = R^nX(t) \qquad (1.33)$$

for n any positive integer. In particular, we
must have

$$X(n) = R^nX(0)$$

where, as before, X(0) gives the initial
population density. The significance of equation
(1.33) is as follows: if $R > 1$, then X(n) will
approach infinity as n becomes arbitrarily large;
if $R = 1$, then X(n) = X(0) for all positive
integer values of n ; and if $R < 1$, X(n)
converges to 0 as n becomes arbitrarily large.
These results may be compared with those
obtained from the Malthusian model of Section 1.1.
In some respects, the parameter r of that model
seems to correspond to R - 1 in its consequences
for the fate of the population. Another angle of

comparison is that of R with e^r , since $e^r = 1$
for $r = 0$. There are good reasons for these
parallels, as will now be shown. Define

$$\Delta X = X(t+1) - X(t) \qquad (1.34)$$

If we take X(t+1) from equation (1.32), we have

$$\Delta X = RX(t) - X(t) = [R-1]X(t) \qquad (1.35)$$

formally parallel to equation (1.5), which had
dx/dt = rx . The other parallel is more difficult
to elucidate. Returning to equation (1.32), let
r = R-1 . We have, following May (1974, p. 27),

$$X(t+1) = (r+1)X(t)$$

giving

$$X(t+n) = (r+1)^nX(t)$$

from equation (1.33). Let us suppose that the
amount of time that passes over n steps is in
fact quite small, letting τ denote this small
amount of time. Given that τ is in fact quite
small, then it would be unreasonable to expect r

to be very large in magnitude. In fact, it would seem reasonable to suppose that $|r| \ll 1$. In this case, we can find a useful approximate expression for $X(t+\tau)$, as

$$X(t+\tau) = (1+r\tau)X(t) + O(r^2\tau) \qquad (1.36)$$

where the expression $O(\ )$ indicates terms of the same magnitude as the argument or smaller. Consider population growth proceeding from some initial value $X(0)$, where the question is what $X(t)$ will be when $t = k\tau$, that is t is k steps of τ units of time from $t = 0$. We must have

$$X(t) \approx (1+r\tau)X([k-1]\tau)$$

$$= (1+r\tau)^2 X([k-2]\tau)$$

$$= (1+r\tau)^k X(0) \qquad (1.37)$$

Now as the population growth process approaches continuity, the step time, τ, approaches zero. Thus we want to find

$$\lim_{\tau \to 0} X(t) = \lim_{\tau \to 0}[(1+r\tau)^k X(0)]$$

$$= \lim_{\tau \to 0}[(1+r\tau)^{t/\tau} X(0)]$$

since $t = k\tau$

$$= X(0)\lim_{\tau \to 0}[(1+r\tau)^{1/\tau}]^t$$

$$= X(0)e^{rt} \qquad (1.38)$$

because

$$\lim_{\epsilon \to 0}[(1+y\epsilon)^{1/\epsilon}] = e^y$$

[This isn't something you were expected to know off-hand.] But equation (1.38) exactly corresponds to equation (1.10), derived as a solution to the Malthusian model. From this, you should be able to see that difference equations have differential equations as limiting approximations, as their time-steps approach infinitesimal values.

Discrete-Time Logistic Model

Given this intimate relationship between models based on differential equations and models based on difference equations, it is logical to try to take models of the first type and convert them into models of the second type. That way we spare ourselves the need to always begin again from first principles in each of our theoretical studies. Consider, as an example, the logistic model of equation (1.15). This model has been considered in the discrete-time context already, by Maynard Smith (1968, pp. 25-26) for example. Proceeding directly from equation (1.15), we write

$$\Delta X = rX(K-X)/K \qquad (1.39)$$

Evidently there are the same two equilibria for model (1.39) that there were for model (1.15):

$$X_1 = 0 \quad \text{and} \quad X_2 = K \qquad (1.40)$$

Consider deviations from these equilibria, as given by

$$X = X_i + x_i \qquad (1.41)$$

with the x_i small, to represent small, or "local", deviations.

For X_1 , we have

$$\Delta x_1 = \Delta(X-X_1) = [X(t+1)-X_1] - [X(t)-X_1]$$
$$= \Delta X = rX(K-X)/K$$
$$= r[X_1+x_1](K-[X_1+x_1])/K$$
$$= rx_1(K-x_1)/K = rx_1 + 0(x_1^2)$$

Since $0(x_1^2)$ terms are very small when X is near X_1 , we examine

$$\Delta x_1 \approx rx_1 \qquad (1.42)$$

Evidently we must have $x_1 > 0$, since we are considering perturbations away from the biological boundary defined by $X = 0$, since negative values of population density are not admissible in the present modelling context. Therefore $\Delta X > 0$ if $r > 0$, from (1.42). We can conclude then that equilibrium X_1 is stable if $r < 0$, and an attractor, while if $r > 0$, X_1 is unstable.

For X_2 , we have

$$\Delta x_2 = r(K+x_2)(K-K-x_2)/K$$
$$\approx -rx_2 \qquad (1.43)$$

as is to be shown in the Exercises. From (1.43), we can infer that, if $r > 0$, $x_2(t+1) < x_2(t)$ for $x_2(t) > 0$ and $x_2(t+1) > x_2(t)$ for $x_2(t) < 0$.

You may then be tempted to conclude that $r > 0$ guarantees convergence of X to X_2 . But this is not correct. If we consider the population's dynamics with various r values, as you are to do in the Exercises, you will see that above a threshold value of r, $X(t)$ does not approach X_2 when $r > 0$.

If r is in the interval $(0,1]$, then near X_2

$$x_2(t+1) = x_2(t) + \Delta x_2$$

$$\approx x_2(t) - rx_2(t)$$

$$= (1-r)x_2(t) \qquad (1.44)$$

Approximation (1.44) can be used to conclude that

$$x_2(t+k) \approx (1-r)^k x_2(t)$$

so that

$$\lim_{k\to\infty} x_2(t+k) \approx x_2(t) \lim_{k\to\infty}[(1-r)^k]$$

Thus the dynamics of $x_2(t)$ depend on the

limiting value of $(1-r)^k$ as k becomes arbitrarily large. If r is in $(0,1]$, then $x_2(t)$ converges to zero, because arbitrarily high

powers of numbers less than 1 in absolute magnitude converge to zero.

If r is in $(1,2)$, then from (1.44)

$$x_2(t+1) \approx (1-r)x_2(t) = -(r-1)x_2(t)$$

which is of opposite sign to $x_2(t)$ because the

value of r is greater than 1. Therefore the trajectory alternates back and forth about X_2.

This is in contrast to the case with r in $(0,1]$, for which (1.44) shows that if $x_2(t) > 0$ then

$x_2(t+1) > 0$, and similarly for $x_2(t) < 0$. Note that

$$|x_2(t+1)|/|x_2(t)| \approx r-1 < 1$$

so that, as before, we can write

$$|x_2(t+k)| \approx (r-1)^k |x_2(t)|$$

Again, as k becomes arbitrarily large the $(r-1)^k$ term converges to zero, because $r-1$ is in $(0,1)$. Therefore we have $|x|$ converging to zero as time passes, so that the population density converges to K .

If $r > 2$, we again have oscillation about X_2 , but

$$|x_2(t+1)| / |x_2(t)| \approx r-1 > 1$$

so that trajectories near X_2 will diverge from it. However, it would be incorrect to assume that such trajectories necessarily proceed to diverge indefinitely from X_2 . In certain cases, though X oscillates about K , it may none the less exhibit a second-order stability called a "limit cycle". Limit cycles arise when trajectories tend toward a specific oscillatory trajectory, like a planetary orbit in a solar system or the Moon's orbit about Earth. In discrete-time models, limit cycles have a specific number of points, of which the simplest are two-point cycles. Figure 1.4 illustrates what a population density trajectory converging to a limit cycle would look like. Sometimes it is possible to solve for such a limit cycle explicitly, in the same way that we have been solving population density dynamical equations for points of equilibrium. In the case of discrete-time dynamical models, we solve for equilibria by considering the values of population density for which $X(t+1) = X(t)$. To find two-point cycles, we solve for $X(t+2) = X(t)$, discounting those solutions for which

$$X(t+2) = X(t+1) = X(t)$$

which are of course equilibrium solutions.

In the present case, from (1.36) we write

$$X(t+1) = X(t)[1 + r(1 - X(t)/K)] \qquad (1.45)$$

so that

$$X(t+2) = X(t)[1+r(1- X(t)/K)] \times \qquad (1.46)$$

$$[1+r\{1-X(t)[1+r(1- X(t)/K)]/K\}]$$

by direct substitution. In order to proceed to our solution for possible two-point limit cycles, let χ represent any such points and let

$$y = r(1 - \chi/K) \qquad (1.47)$$

Then $X(t+2) = X(t)$ when

$$0 = \chi y[2 - r\chi/K + r\{1 - \chi(1+y)/K\}] \qquad (1.48)$$

as is to be shown in the Exercises. There are two trivial solutions to (1.48) which are readily discounted: $\chi = 0$ and $y = 0$, corresponding to equilibria X_1 and X_2 , respectively. The other solutions arise when

$$0 = 2 - r\chi/K + r\{1 - \chi(1+y)/K\} \qquad (1.49)$$

Obviously only the roots of (1.48) which are also roots of (1.49) are possible limit cycle points. [A root is a solution to an algebraic equation which normally has been set-up so that the LHS is zero. In this case, we are solving for χ values, of course.] To investigate equation (1.49) further, we need to express it in terms of only one unknown variable. In the end we want to know χ values,

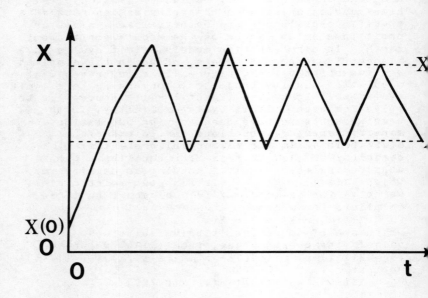

Figure 1.4. A population density trajectory converging on a two-point limit cycle in a discrete-time model of population growth.

but for now it will suffice to consider y solutions, since there is a simple relationship between y and χ. Using (1.47), (1.49) becomes

$$0 = y^2 + y[2 - r] + 2 - r \qquad (1.50)$$

as is to be shown in the Exercises. From the formulae

$$[-b + \sqrt{b^2-4ac}]/2a \quad \text{and} \quad [-b - \sqrt{b^2-4ac}]/2a \quad \text{for}$$

the roots of

$$ax^2 + bx + c = 0$$

for equation (1.50) we find the roots

$$y_+ = [r - 2 + \sqrt{r^2-4}]/2$$

and $\qquad (1.51)$

$$y_- = [r - 2 - \sqrt{r^2-4}]/2$$

as is to be shown in the Exercises. Using equation (1.47), we then find the roots in their final form:

$$X_+ = K[1 - \{r - 2 + \sqrt{r^2-4}\}/2r]$$

and $\qquad (1.52)$

$$X_- = K[1 - \{r - 2 - \sqrt{r^2-4}\}/2r]$$

However, finding a pair of roots like X_+ and X_- is no guarantee that they define an

asymptotically stable limit cycle, the moving equivalent of an attractor point. To assess the stability of such two-point limit cycles, the difference equation must be iterated twice, and then assessed for the behaviour of the system upon perturbation from both of the two roots. More formally, if we have a difference equation of the form

$$X(t+1) = f[X(t)] \qquad (1.53)$$

then what is of interest is the local stability of the dynamics defined by a new equation given by

$$Y(t+1) = f\{f[Y(t)]\} \qquad (1.54)$$

where $Y(t+1)$ corresponds to $X(t+2)$. For discrete-time dynamical systems like (1.54), the local asymptotic stability of equilibria, say X^*, requires

$$|df[X^*]/dX| < 1 \qquad (1.55)$$

In the case of model (1.54), the appropriate criterion remains the same, except that the relevant derivative is

$$df\{f[X^*]\}/dX = df\{f[X^*]\}/df[X] \times df[X^*]/dX \qquad (1.56)$$

But if X^* is X_+ in the case of a two-point

limit cycle, then $f[X^*]$ is in fact X_-. Therefore equation (1.56) becomes

$$df\{f[X^*]\}/dX = df[X_+]/dX \times df[X_+]/dX \qquad (1.57)$$

whichever of the two roots is chosen as X^*. This

result is readily generalized (León, 1975), so that for limit cycles involving T points, local asymptotic stability requires

$$\left| \pi_i \, df[X_i]/dX \right| < 1 \qquad (1.58)$$

where the symbol π_i indicates the product of all the derivatives from $i = 1$ to $i = T$ and the X_i are the points of the limit cycle.

In the case of the discrete-time logistic model given by equation (1.39),

$$f[X(t)] = X(t)\{1 + r[K-X(t)]/K\}$$

so that

$$df[X]/dX = 1 + r - 2rX/K$$

and local asymptotic stability of the limit cycle therefore requires

$$\left| (1 + r - 2r\chi_+/K)(1 + r - 2r\chi_-/K) \right| < 1$$

which simplifies to

$$\left| 5 - r^2 \right| < 1 \qquad (1.59)$$

as is to be shown in the Exercises. Result (1.59) shows that for r in the interval $(2, \sqrt{6})$, the two-point limit cycle defined by the points of (1.52) is locally asymptotically stable.

For values of r greater than $\sqrt{6}$, the two-point limit cycle is unstable, and more complicated attractors arise, as you are to explore numerically in the Exercises. The potential diversity of behaviour exhibited by such ostensibly simple equations includes: (i) more complex limit cycles, of longer period; (ii) completely chaotic behaviour, with fluctuations of no periodicity, but no extinction of the population, referred to in terms of "strange attractors"; and (iii) rapid extinction of the population due to overshoot of the carrying-capacity, with X crashing to negative values. This diversity of behaviour is discussed in May and Oster (1976), as well as many other recent publications in the field.

General Autonomous Models

As for Figure 1.3 of Section 1.2, we can have arbitrary curves giving the reproductive rate as a function of X :

$$X(t+1) = R[X(t)]X(t) \qquad (1.60)$$

At equilibria, say X_i,

$$X_i = R[X_i]X_i$$

so that we must have

$$R[X_i] = 1 \qquad (1.61)$$

Thus we have equilibria at intersections of $R[X]$ and $Y = 1$, as in Figure 1.5. Consider the interval (X_1, X_3). For $X(t)$ near X_1, with

$X(t) > X_1$, $X(t+1) > X(t)$ from the Figure. For

$X(t)$ near X_3, with $X(t) < X_3$, $X(t+1) < X(t)$.

Thus both X_1 and X_3 are unstable equilibria,

which the population trajectory at first moves away from in their neighbourhood. Near X_2, for

$X(t) < X_2$, $X(t+1) > X(t)$, while for $X(t) > X_2$,

$X(t+1) < X(t)$. Thus trajectories starting near X_2

move toward it. So, for $X(t)$ in (X_1, X_3), the trajectory proceeds toward X_2.

However, $X(t)$ may overshoot, as we have already found in the case of the discrete-time logistic model. This can be shown formally by generalizing the analysis of the logistic model to model (1.60). We write, following Maynard Smith (1968, pp.21-23),

$$X(t) = X_2 + x \qquad (1.62)$$

Again, in the neighbourhood of X_2 we have the following Taylor expansion

$$\begin{aligned} R[X] &= R[X_2] + (X-X_2)dR[X_2]/dX \\ &\quad + \{(X-X_2)^2/2\}d^2R[X_2]/dX^2 \\ &\quad + \ . \ . \\ &= 1 + xdR[X_2]/dX + O(x^2) \\ &\approx 1 + xdR[X_2]/dX \qquad (1.63) \end{aligned}$$

This is of course just the same sort of approach that we took with model (1.23), where equation (1.25) was the analogue of equation (1.63). Given the plot of $R[X]$ in Figure 1.5, we must have $dR[X_2]/dX < 0$.

From (1.63) and (1.60), near X_2 we have
$$\begin{aligned} X(t+1) &= R[X(t)]X(t) \\ &\approx \{1 + x(t)dR[X_2]/dX\}X(t) \end{aligned}$$
so that

$$X_2 + x(t+1) \approx \{1 + x(t)dR[X_2]/dX\}\{X_2 + x(t)\}$$

$$\approx X_2 + x(t) + X_2 x(t)dR[X_2]/dX + O(x^2)$$

giving

$$x(t+1) \approx x(t)\{1 + X_2 dR[X_2]/dX\}$$

and thus

$$\Delta x \approx \{X_2 dR[X_2]/dX\}x \qquad (1.64)$$

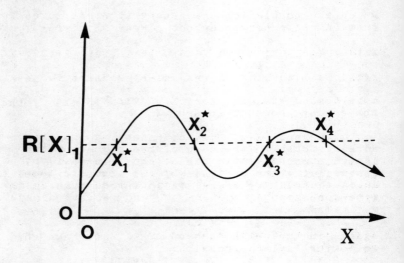

Figure 1.5. A curve depicting the dependence of population growth on density for discrete-time model (1.60), given by $R[X]$. The X_i

are the population density equilibria for which $R[X] = 1$, as indicated by the intersection of the curve for $R[X]$ with $Y = 1$.

Approximation (1.64) shows that $\Delta x < 0$. If we let

$$b = -X_2 dR[X_2]/dX \qquad (1.65)$$

then it is clear from the same sort of argument as the one we used in analyzing the behaviour of the logistic model near K that: (i) b < 0 gives divergence from X_2 ; (ii) b = 0 indicates that

higher-order analysis is required; (iii) b in (0,1] results in the population density approaching X_2 without oscillation; (iv) b in

(1,2) results in diminishing oscillations as the population density converges on X_2 ; (v) b > 2

gives divergent oscillations about X_2 , with the

possibility of convergence to limit cycles. Thus the kind of findings that we obtained for the logistic model are in no sense peculiar; they arise generally when difference equations are used for models of population growth.

Density-Independent Nonautonomous Models

Again we have the same dichotomy as that discussed in Section 1.2, between models which are autonomous and those which are not. So far we have been discussing discrete-time models which are strictly autonomous only. Now we turn to models which have time-dependent parameters, but no density-dependent parameters.

Following Hoppensteadt (1976, pp.5-7), the basic growth equation is

$$X(t+1) = R_t X(t) \qquad (1.66)$$

where R_t can change each generation. The

solution to model (1.66) is obtainable by induction as

$$X(t) = R_{t-1}R_{t-2} \cdot \cdot R_1 R_0 X(0) \qquad (1.67)$$

although this isn't particularly informative. If we define

$$\prod_{k=0}^{t-1} R_k = R_{t-1}R_{t-2} \cdot \cdot \cdot R_1 R_0$$

which is conventional mathematical notation, then we can write (1.67) as

$$X(t) = [\prod_{k=0}^{t-1} R_k]X(0) \qquad (1.68)$$

Equation (1.68) does yield to analysis to some extent. Taking natural logarithms of both sides, we

find
$$\ln X(t) = \ln \{[\pi_k R_k]X(0)\}$$
$$= \ln [\pi_k R_k] + \ln X(0)$$
$$= \sum_{k=0}^{t-1}\ln R_k + \ln X(0) \qquad (1.69)$$

where we now have a sum of ln functions because
$$\ln [abc] = \ln a + \ln b + \ln c$$
From equation (1.69), we can gain some idea of the average density of the population after a great deal of time has elapsed as follows:

$$\lim_{t \to \infty}\{t^{-1}\ln X(t)\} = \lim_{t \to \infty}\{t^{-1}\Sigma_k \ln R_k + t^{-1}\ln X(0)\}$$
$$= \lim_{t \to \infty}\{t^{-1}\Sigma_k \ln R_k\}$$
$$= \ln R_m \qquad (1.70)$$

thereby implicitly defining R_m . [The

(1/t)ln X(0) term disappeared because $\lim_{x \to \infty} c/x = 0$, when c is a constant.] Thus
$$\lim_{t \to \infty} \ln X(t) \approx t \ln R_m$$

so that
$$\ln X(t) \approx \ln [(R_m)^t]$$
and
$$X(t) \approx (R_m)^t \qquad (1.71)$$

where the t on the RHS is a conventional exponent, rather than a superscript. From a practical standpoint, equation (1.71) doesn't have a great deal of value as an absolute prediction, since one has to wait till an arbitrarily large amount of time has passed. However, there are certain features of (1.71) which may be somewhat enlightening. In particular, if $R_m < < 1$, then

X(t) will tend to approach 0 , while if $R_m > > 1$

then X(t) will tend to become large. [This may be compared with the similar findings for the density-independent nonautonomous model discussed in Section 1.2.] Otherwise, we need to have some information about the factors determining the R_i in order to say more about the population trajectory.

Time-Lag Models
One of the major deficiencies of density-dependent

autonomous models of single-species population growth is that most of them implicitly presume that the response of regulating species, such as predators, parasites, competitors and pathogens, is instantaneous. While this may be essentially the case for large-brained predators, such as birds, or the numerical response of pathogens with extremely short generation times, like some viruses, it does not hold in general. Some predators have generation times longer than their prey together with little important functional response to changes in abundance. The result is that the effective regulatory response is of about the same rapidity as the population growth process itself, rather than an instantaneous function of it. One way of dealing with this problem, of course, is to model the density-dependent interactions between such species explicitly. Another way to deal with it is to incorporate time-lags into the equations for population growth.

Perhaps the simplest introduction to the effects of time-lags on single-species population growth models is that afforded by difference-equation models with 1-step time-lags (Maynard Smith, 1968, pp. 23-25). Consider the following model:

$$X(t+1) \quad = \quad R[X(t-1)]X(t) \qquad (1.72)$$

In this model, the rate of population growth at time t depends on the population's density at time $t-1$. Thus this model presumes that there is in fact no instantaneous regulatory response; all such responses being confined to the next time-step. This isn't as unreasonable as it might seem. Univoltine parasitoids of univoltine herbivorous insect hosts can give rise to dynamics very much like this. [The biology of these organisms will be discussed further in Chapter 3.]

Given equation (1.72), we can also write

$$X(t+2) \quad = \quad R[X(t)]X(t+1) \qquad (1.73)$$

The population's dynamics about equilibria remain accessible to the same sort of treatment that we have given before. We define deviations away from any equilibria as

$$X(t) \quad = \quad X* + x(t) \qquad (1.74)$$

with $R[X*] = 1$, just as before. Again, we approximate $R[X]$ using the initial terms of a Taylor series expansion:

$$R[X(t)] \quad \approx \quad 1 \quad + \quad x(t)dR[X]/dx\big|_{X=X*}$$

If we set

$$b \quad = \quad -X*dR[X*]/dX$$

as in equation (1.65), then we can write

$$R[X(t)] \approx 1 - bx(t)/X* \qquad (1.75)$$

Substituting (1.74) and (1.75) into equation (1.73) gives

$$X*+x(t+2) \approx [X*+x(t+1)][1-bx(t)/X*] \qquad (1.76)$$

and thus, rearranging approximation (1.76) and dropping $O[x^2]$ terms,

$$x(t+2) \approx x(t+1) - bx(t)$$

or

$$x(t+2) - x(t+1) + bx(t) \approx 0 \qquad (1.77)$$

We now have a model for the dynamics of the population near arbitrary equilibria defined by the roots of $R[X] = 1$. What we want now is to discover how the solutions of (1.77), and thus population dynamics near equilibria, depend on b since this is the only biologically interpretable parameter in the model. Thus we need to solve (1.77) in terms of a formula relating $x(t)$ to t and b.

There are no universal formulae for solving difference or differential equations when variables are not separable. The usual method is to try standard possible solutions which have been found to work for similar equations, where such standard solutions have free parameters which can be used to fit the particular case at hand. [You may be more familiar with this solution-strategy by the name "trial-and-error".] The standard solution formula for equations like (1.77) is

$$x(t) = C\lambda^t$$

where C and λ are "chosen to fit" the equation. If we try this type of solution out in (1.77), we must have C and λ such that

$$C\lambda^{t+2} - C\lambda^{t+1} + bC\lambda^t = 0$$

or

$$\lambda^2 - \lambda + b = 0 \qquad (1.78)$$

once both sides have been divided by $C\lambda^t$. This is a quadratic formula with roots

$$\lambda_{\pm} = 1/2 \pm \sqrt{1-4b}/2 \qquad (1.79)$$

In difference and differential equation problems of this kind, for which there are usually two such roots as elements in possible solutions, the actual solution must be a linear combination of the solutions afforded by the two roots. Thus we seek a solution of the form

$$x(t) = D\lambda_+{}^t + E\lambda_-{}^t \qquad (1.80)$$

where the D and E factors are chosen to fit the initial conditions.

One problem which may arise is that 1-4b may be less than zero, in which case the roots from (1.79) may have roots of negative numbers. Such roots are "complex" numbers, in that they involve everyday ("real") factors along with factors which can be expressed as a real factor times $\sqrt{-1}$, the latter normally abbreviated to "i" . When 1-4b is less than zero, $\sqrt{1-4b}$ = $i\sqrt{4b-1}$. If you do not already know, you are probably wondering what powers of complex numbers, like those which may arise in equation (1.80), look like, particularly when they are supposed to result in a conventional real number, x , as a function of time. One way to acquire some insight into the effects of complex exponents in equations like (1.80) is to consider the exponential function when it has an argument

involving i . The definition of e^x is as follows:

$$e^x = 1 + x + x^2/2! + x^3/3! + x^4/4! + . .$$
when x is a conventional Real number. Now

consider e^{ix} , using the same definition:

$$e^{ix} = 1 + ix - x^2/2! - ix^3/3! + x^4/4! + . .$$
This expression can be reorganized into those terms involving i and those not involving i , as follows:

$$e^{ix} = [1 -x^2/2! +x^4/4! ..] + i[x -x^3/3! + ..]$$
In fact, the expressions in the brackets are cos x and sin x respectively. Thus we have

$$e^{ix} = \cos x + i \sin x$$

In fact, if λ^j is complex, then it can be

expressed as $\phi e^{i\theta}$ for appropriate ϕ and θ . The puzzle that may remain in your mind is how x(t) , which is a perturbation of a population density, can have a value which involves i . The solution to this puzzle lies in choosing D and E parameters for equation (1.80) which result in cancellation of those terms involving i . When this is all done correctly, and you can be spared the details here, the solution to (1.80) has the form

$$x(t) = (\sqrt{b})^t[\alpha\cos(t\theta) + \beta\sin(t\theta)] \qquad (1.81)$$

where $\theta = \cos^{-1}([2\sqrt{b}]^{-1})$ and α and β are arbitrary real-valued constants. [See Maynard Smith (1968, Appendix 4) for the details of how this equation is derived.]

Let us return to equation (1.80) itself, and the roots given in (1.79). For b in the interval $(0, 1/4)$, we can use (1.80) directly, since the roots given by (1.79) are real. Irrespective of D and E, we have $x(t)$ converging to 0 as time increases indefinitely, because both λ_+ and λ_-

are in the unit interval, because $\sqrt{1-4b}$ is in $(0,1)$ so that when half this quantity is added to $1/2$, the resulting value does not exceed 1. Thus the largest magnitude root does not exceed 1 in magnitude. If $b < 0$, we have

$$\lambda_+ = 1/2 + \sqrt{1-4b}/2 > 1/2 + \sqrt{1}/2 = 1$$

which ensures that the $D\lambda_+$ term of (1.80)

explodes as t goes to ∞, giving instability of X^* for arbitrary $x(0)$. If $b > 1/4$, then we must use formula (1.81), because the roots given by (1.79) are now complex. Evidently, the cosine and sine functions do not converge to zero or expand to infinity as t goes to ∞. They just generate oscillations. Convergence [x(t) going to 0] or

divergence [x(t) going to ∞] depend on the $(\sqrt{b})^t$ factor. In particular, for $b < 1$, we evidently have this factor converging on 0, and thus $x(t)$

converging on 0. For $b > 1$, $(\sqrt{b})^t$ goes to ∞ as t goes to ∞. Evidently either case, convergence or divergence, involves oscillation. You may have noticed that we have not considered the "boundary cases", such as $b = 1/4$. Such cases are mathematically tricky and, in any case, fall into the category of the biologically implausible, in that we do not expect to find biological systems perched precisely on any particular parameter value. Therefore you have been spared them.

The effects of time-lags on the stability of population densities are readily illustrated by comparing these results with those for population dynamics in discrete-time without time-lags. Then Table 1.1 shows that the requirements for convergence are more stringent when there are time-lags than when there are no such lags. This suggests that time-lags may destabilize ecosystems, under some circumstances at least.

Table 1.1. Effect of the parameter b on the local population dynamics about the equilibria of discrete-time population growth models.

Local Dynamics About Equilibria	No Time-Lag	1-Step Lag
divergence	$b < 0$	$b < 0$
non-oscillatory convergence	$0 < b < 1$	$0 < b < 1/4$
oscillatory convergence	$1 < b < 2$	$1/4 < b < 1$
oscillatory divergence	$2 < b$	$1 < b$

1.4 MODELS WITH AGE-STRUCTURE

One of the most important weaknesses of the models that we have treated so far is that they assume implicitly that every individual has equal probabilities of survival and reproduction. Evidently such a model is radically unrealistic for multicellular organisms, virtually all of which must pass through an immature stage between birth and reproduction. Thus age-structured models of population dynamics have become an important part of mathematical ecology.

Discrete-Time: The Leslie Matrix
In discrete-time, the now classical model is that based on the "Leslie matrix" (Lewis, 1942; Leslie, 1945, 1948; Pielou, 1977, pp. 41-57). This model assumes: (a) that both time and biological age advance in discrete steps; (b) that, within a given "age-class" [all those organisms having the same age], birth and death rates are constant; and (c) that the population does not exhibit ecological differences between the sexes, perhaps because the organism is hermaphroditic, or the population growth process depends only on female age-structure, fertile males never being a limiting factor. [For a discussion of various age-structured population growth models which deviate from these restrictive assumptions, see Charlesworth (1980,

Chapter 1).]

Let us develop the population growth model which is associated with the Leslie matrix. Firstly, we specify the variable. Let $\underline{n}(t)$ be a vector giving the age-structured composition of the population,

$$\underline{n}(t) = \begin{bmatrix} n_0(t) \\ n_1(t) \\ \cdot \\ \cdot \\ n_i(t) \\ \cdot \\ \cdot \\ n_d(t) \end{bmatrix}$$

where i is the i-th age class, arranged according to chronology, with the 0-th age class being the "newborns" and the d-th age class being the last age-class which reproduces. One example of such an age-class system is the conventional human designation of age in years, where individuals aged i years + j days, with j < 365 [or 366 in the case of leap years], being designated as age i .

Secondly, we associate the variables of interest with a model giving their rates of change, as before. Let F_i be defined as the number of new

"reproductive progeny" [since we may be keeping track of females only] produced in one time-step [such as one year in the case of a human population] by a reproductive organism in age-class i , surviving into the next time interval. Let P_i be the probability that a reproductive organism aged i at time t will remain a surviving reproductive at time t+1 and age x+1 . With these definitions, we have the following discrete-time dynamical system:

$$n_0(t+1) = F_0 n_0(t) + \ldots + F_d n_d(t) \qquad (1.82.0)$$

$$n_1(t+1) = P_0 n_0(t) \qquad (1.82.1)$$

$$n_2(t+1) = P_1 n_1(t) \qquad (1.82.2)$$

$$\cdot \qquad \qquad \cdot$$
$$\cdot \qquad \qquad \cdot$$

$$n_d(t+1) = P_{d-1} n_{d-1}(t) \qquad (1.82.d)$$

where d is the last age of reproduction. (Note that $1 - P_i$ gives the probability of either death

or final cessation of reproduction in this model.)

A more convenient way of writing the $d+1$ equations of (1.82) is to replace them by a matrix equation:

$$\underline{n}(t+1) = M\underline{n}(t) \qquad (1.83)$$

where

$$M = \begin{bmatrix} F_0 & F_1 & \cdot & \cdot & \cdot & \cdot & F_{d-1} & F_d \\ P_0 & 0 & \cdot & \cdot & \cdot & \cdot & 0 & 0 \\ 0 & P_1 & 0 & \cdot & \cdot & \cdot & 0 & 0 \\ 0 & 0 & P_2 & \cdot & \cdot & \cdot & \cdot & \cdot \\ \cdot & \cdot & \cdot & \cdot & & & & \cdot \\ \cdot & \cdot & & & \cdot & & & \cdot \\ 0 & 0 & & & & & P_{d-1} & \cdot \end{bmatrix} \qquad (1.84)$$

where the " . . . " notation indicates repetition of like elements along the column, row, or diagonal so labelled. From equation (1.83), a solution to the model may be found straightforwardly:

$$\underline{n}(t+2) = M\underline{n}(t+1) = M \times M\underline{n}(t) = M^2\underline{n}(t)$$

so that we can conclude, by induction, that

$$\underline{n}(t) = M^t\underline{n}(0) \qquad (1.85)$$

Now we have the task of trying to extract some information from "solution" (1.85), which by itself tells us nothing. An important feature of the definition of M given by equation (1.84) is that it cannot contain negative entries. Therefore all the mathematical findings for non-negative matrices (e.g., Pullman, 1976) can be brought to bear on the question of what equation (1.85) means. The piece of mathematics which turns out to be most important is the Perron-Frobenius Theorem (Pielou, 1977, p.44) which allows us to conclude that there is a

stable age-distribution $\underline{n}^*$ such that $\lim_{t \to \infty} M^t\underline{n}(0)$

is proportional to $\underline{n}^*$ providing there are no bizarre periodicities in the numerical values of the F_i (Charlesworth, 1980, pp.25-29). From this,

as $t \to \infty$, we must have

$$[M^{t+1}\underline{n}(0)]/[M^t\underline{n}(0)] = \lambda^* \qquad (1.86)$$

where λ^* is a real-valued [i.e., not complex] scalar [i.e., not a vector] parameter, and so we can write

$$\underline{n}(t+s) = (\lambda^*)^s\underline{n}^* \qquad (1.87)$$

for t sufficiently large. [Numerically, "sufficiently large" is often not more than $t > 50$.] The key components in these equations, $\underline{n}^*$ and λ^* , are directly calculable from M , being known as its "dominant eigenvector" and "dominant eigenvalue", respectively. The latter is obtained from the largest real-valued root of the equation

$$|M - \lambda I| = 0 \qquad (1.88)$$

where I is the identity matrix

$$I = \begin{bmatrix} 1 & 0 & . & . & . & . & 0 \\ 0 & 1 & 0 & . & . & . & 0 \\ 0 & 0 & 1 & 0 & . & . & 0 \\ . & . & & & & . & . \\ . & . & & & & . & . \\ 0 & 0 & . & . & . & 1 & 0 \\ 0 & 0 & . & . & . & 0 & 1 \end{bmatrix}$$

and $|\ \ |$ indicates the operation of finding the determinant of the matrix between the two vertical lines. [We will be discussing the nature of determinants in detail in Chapter 2. For now, it is enough to know that a determinant is a single scalar variable associated with any given matrix. If you want to know more at this point, most elementary texts on linear algebra or matrices explain determinants.] Once $\lambda*$ has been obtained, $\underline{n}*$ is found from the solution to the equation

$$M\underline{n}* = \lambda*\underline{n}* \qquad (1.89)$$

Equation (1.88) should already be fairly well-known to you in a different guise. In full, this equation is written as

$$\begin{bmatrix} F_0-\lambda & F_1 & F_2 & . & . & . & . & F_m \\ P_0 & -\lambda & 0 & . & . & . & . & 0 \\ 0 & P_1 & . & . & . & . & . & 0 \\ . & & . & . & . & . & . & . \\ . & & & . & & & & . \\ . & & & & . & -\lambda & & 0 \\ 0 & . & . & . & . & 0 & P_{d-1} & -\lambda \end{bmatrix} = 0$$

By means of tedious algebra or matrix transformation (e.g., Pielou, 1977, p.46), it can be shown that equation (1.88) can be written as

$$\lambda^{d+1} - F_0\lambda^d - P_0F_1\lambda^{d-1} - P_0P_1F_2\lambda^{d-2} - . .$$
$$- (P_0P_1P_2 . . P_{d-1})F_d = 0 \qquad (1.90)$$

Let
$$l_x = \prod_{k=0}^{x-1} P_k \quad , \quad l_0 = 1 \qquad (1.91.1)$$
and
$$m_x = F_x \qquad (1.91.2)$$

where x is now age, not density, and ranges over the whole number values from 0 to d . Then we can write (1.90) as

$$\lambda^{d+1} - l_0 m_0 \lambda^d - l_1 m_1 \lambda^{d-1} \cdots - l_d m_d = 0$$
or
$$\lambda^{d+1} = \Sigma_x \, l_x m_x \lambda^{d-x}$$

or, having divided by λ^{d+1},

$$1 = \Sigma_x \, l_x m_x \lambda^{-x-1} \qquad (1.92)$$

which is the famous Euler-Lotka characteristic equation to be found in so many introductory ecology texts.

The interpretation of λ^* is that it represents the asymptotic population growth rate, which is obvious enough from equation (1.89). From this equation, it is also clear that the population growth process depends on λ^* in much the same way that it depended on R in the density-independent autonomous model of population growth in discrete-time: λ^* in [0,1) gives population extinction; $\lambda^* = 1$ gives stable population densities; and $\lambda^* > 1$ gives indefinite expansion of population size. Thus λ^* is a readily interpretable population growth parameter. It isn't easy to find, ·numerically, but you needn't worry about that here. What we have shown is that population age-structure doesn't necessarily give population growth patterns which are that different from those in populations without age-structure, at least after the stable age-distribution has been reached. As Charlesworth (1980, Chapter 1) discusses, such stable age-distributions are normally achieved quite quickly.

What we have discussed here has been confined to cases without density-dependent effects upon survival or reproduction. Models of population growth incorporating both age-structure and density-dependence have been developed (cf. Charlesworth, 1980), but they require fairly advanced mathematics to analyse, and so are beyond our scope here.

Continuous-Time Models

As in our discussion of models without age-structure, we consider the continuous-time analogue of the discrete-time model of age-structured population growth. Following Pielou (1977, pp.58-61) or Roughgarden (1979, pp.328-329), we begin by assuming that the population has achieved a stable age-distribution, with the average birth and death rates remaining constant. Therefore we can assume the validity of equation (1.4) here, with

$$dN/dt = (b-d)N = rN$$

We again take x to represent age, rather than population density, and take as our known variables: (i) $l(x)$, the probability of survival to age x, from birth; (ii) $m(x)$, the mean number of offspring born to organisms in the age-interval $(x,x+dx)$; (iii) $C(x)dx$, the proportion of organisms in the age-interval $(x,x+dx)$ in the stable age-distribution; and (iv) $B(t)dx$, the number of new organisms produced in a time-interval of length dx at time t.

Once the population has achieved the stable age-distribution, at time t, the number of individual organisms within the age-interval $(x,x+dx)$ is equal to $N(t)C(x)dx$, by definition. It is also equal to $l(x)B(t-x)dx$, because these individuals must have been born at time $t-x$ and must have survived to age x. Thus we have

$$N(t)C(x)dx = l(x)B(t-x)dx \qquad (1.93)$$

From the stable age-distribution assumption, we must have $b = B(t)/N(t)$, for all t, and $N(t)$ must be given by the analogue of (1.10):

$$N(t) = N(t-x)e^{rx}$$

from equation (1.4). Thus we can write

$$B(t-x) = bN(t-x) = bN(t)e^{-rx} \qquad (1.94)$$

From equations (1.93) and (1.94)

$$C(x) = l(x)B(t-x)/N(t) = l(x)be^{-rx} \qquad (1.95)$$

From our definitions, we can also write

$$B(t) = \int_{0}^{\infty} N(t)C(x)m(x)dx$$

$$= \int N(t)[l(x)be^{-rx}]m(x)dx \qquad (1.96)$$

If we extract all terms which do not have x as an argument from the integral, we find

$$(1/b)B(t)/N(t) = \int e^{-rx}l(x)m(x)dx$$

But $b = B(t)/N(t)$, so that we have the

continuous-time version of the Euler-Lotka characteristic equation:

$$1 = \int_0^\infty e^{-rx} l(x)m(x)dx \qquad (1.97)$$

Given r from equation (1.97), it only remains to calculate $C(x)$ to have a complete characterization of the population growth process once a stable age-distribution has been achieved. Since $C(x)$ is a proportion, we require

$$C(x) = B(t-x)l(x)/N(t)$$
$$= B(t-x)l(x)/\int B(t-u)l(u)du \qquad (1.98)$$

Now the assumptions that equation (1.4) holds and the age-distribution is stable give

$$B(t-x) = e^{r(t-x)}B(0)$$

since $B(t)$ gives the size of the $(0,dx)$ age-class at time t. Therefore equation (1.98) is equivalent to

$$C(x) = e^{r(t-x)}B(0)l(x)/\int e^{r(t-u)}B(0)l(u)du$$
$$= e^{-rx}l(x)/\int e^{-ru}l(u)du \qquad (1.99)$$

The treatment that we have just gone through of continuous-time age-structured populations was as simple as possible. It was not shown how populations converge to stable age-distributions, nor why the exponential solution to equation (1.4) could be assumed. These and other advanced topics in the theory of population growth with age-structure are reviewed by Charlesworth (1980, Chapter 1) and Roughgarden (1979, Chapter 18).

1.5 EXERCISES

Elementary

1. Derive equation (1.19) from equations (1.17) and (1.18).

2. How long does it take for population density to double with model (1.5)?

3. For model (1.15), show that $dx/dt > 0$ for $K > x > 0$ and $dx/dt < 0$ for $x > K$.

4. Solve equation (1.28) for $r(t) = 1/t$, $t > 1$.

5. Analyse $dx/dt = r(x)x$ with $r(x) < 0$ for x in $[0,x_1)$, $r(x) > 0$ for x in (x_1,x_2), and

$r(x) < 0$ for x in (x_2, ∞) .

6. Derive equation (1.43).

7. Derive equations (1.48), (1.50), and (1.51).

Intermediate

8. At what density does population growth decelerate with the continuous-time logistic model and at what time is this density reached?

9. Analyse the asymptotic behaviour of populations subject to the density-dependent pattern of population growth indicated in Figure 1.3.

10. If $dx/dt = e^{-t}x$, what is the asymptotic population density?

11. If $X(0) = 10$, $K = 100$, and model (1.39) holds, explore the model's behaviour analytically, when possible, and numerically, by repeated iteration of the difference equation to estimate the next generation's population density, when r takes on the following values: (a) 0.5 , (b) 1.5 , (c) 2.9, (d) 4.0 .

12. Show that limit cycle stability for the discrete-time logistic model requires $|5-r^2| < 1$.

Advanced

13. Analyse two-point limit cycles for the discrete-time logistic model with an additional one-step time lag.

Chapter Two

COMPETITION

In the preceding chapter, we treated the population growth of a single species, not as if the species were inhabiting an ecological vacuum, but as if all effects of other species could be incorporated in time-dependent or density-dependent functions determining the net reproductive rate of individuals of that species. Now we turn to cases in which some of the other species are removed from these functions and explicitly represented in the model. In particular, if we label two species "1" and "2" , then we will give their numbers as N_1 and N_2 , respectively, and so on, with similar subscripts always used to denote the parameters affecting the dynamics of these two species. We define the densities of the two species by

$$x_i = N_i/A \qquad (2.1)$$

with i = 1 and 2 as well as A as before. This equation is evidently analogous to equation (1.1). These will be the two variables of interest to us in our studies of competition. [Others, such as species biomass, could have been chosen instead.]

Any two species can have a variety of interactions with one another. One may be a predator, the other its prey. They could be commensal, or symbiotic, species, facilitating each other's reproduction. Alternatively, they could impede one another's reproduction, neither benefiting, in which case we consider them competitors. [The concrete mechanisms underlying such competition can be quite diverse, from shared and limited food resources, fouling each other's environment, and so on.] We begin with the case of competition, because it is often simplest mathematically.

Implicit in this definition of competition is
the assumption of density-dependence between
species. Thus we assume that as one species
increases in density, the net reproductive rate of
individuals of the other species is progressively
reduced, and <u>vice versa</u>. In formal terms, we might
have the following equation for the population
growth of species i :

$$dx_i/dt = f_i(x_i, x_j)x_i \qquad (2.2)$$

where j is not equal to i . Then if i and j
are only competitors, we must have

$$\partial f_i(x_i, x_j)/\partial x_j < 0$$

for all values of i and j , providing $i \neq j$.
[If you don't know the $\partial y/\partial x$ notation, it will be
explained shortly.] In mathematical terms, this is
the essential meaning of the ecological concept of
competition.

2.1 LOTKA-VOLTERRA MODELS: SPECIAL CASES

In the 1920's, two mathematicians, Vito Volterra
and A.J. Lotka, proposed a variety of elementary
mathematical models of population dynamics (Scudo
and Ziegler, 1978). In particular, they proposed
competition models with autonomous (recall that
that means models with parameters which do not
depend on time) and linear functions relating the
reproductive rate of individual organisms to the
densities of their own species and competitor
species. These are therefore now known as
"Lotka-Volterra competition models". Here we will
be concerned initially with two-species versions of
these models.

No Carrying Capacities
We begin by assuming that neither species has any
"self-regulatory" effects on its own density. That
is, each species grows according to the Malthusian
model in the absence of the other species. Since
we are also assuming linearity of reproductive rate
functions, this assumption directly leads to
equations with reproductive rate functions for each
species as follows:

$$r_i(x_i, x_j) = \alpha_i - \beta_i x_j \qquad (2.3)$$

This gives us the following equations for the
population dynamics of the two species:

$$dx_1/dt = x_1(\alpha_1 - \beta_1 x_2) \qquad (2.4.1)$$

$$dx_2/dt = x_2(\alpha_2 - \beta_2 x_1) \qquad (2.4.2)$$

The α_i's are referred to as the species "growth rates" while the β_i's are referred to as the "competition coefficients". The latter are necessarily positive in cases of competition.

Having decided upon our model, we proceed to analyse it (cf. Freedman, 1980, pp.144-145). First, consider equation (2.4.1) with $\alpha_1 < 0$.

Irrespective of x_2 , $dx_1/dt < 0$, and thus $x_1(t)$ goes to zero as t goes to ∞ . In the special case $\alpha_1 = 0$, $dx_1/dt < 0$ so long as $x_1(t) > 0$ and $x_2(t) > 0$, so that again $x_1(t)$ goes to zero as t goes to ∞ , so long as $x_1(0) > 0$ and $x_2(0) > 0$. You may be wondering what might happen if there were some time, say τ , at which $x_2(\tau) = 0$, even though $x_2(0) > 0$. In fact, such cases cannot arise. In a continuous-time model, as we discussed in Chapter One, a variable which starts off greater than zero can never actually equal zero, it can only approach it asymptotically, after a great deal of time has elapsed. [If either $x_i(0) = 0$, then the model is not one of competition, and we dismiss it from consideration.] Thus if $\alpha_1 \leq 0$, we know that $x_1(t)$ goes to zero in all cases of scientific interest. But what of $x_2(t)$ when $\alpha_1 \leq 0$? If $\alpha_2 \leq 0$, then $x_2(t)$ asymptotically approaches zero, just like $x_1(t)$.

If $\alpha_2 > 0$ and $x_2(0) > 0$, then $dx_2/dt > 0$ when

$$x_2(\alpha_2 - \beta_2 x_1) > 0$$

which requires

$$\alpha_2 > \beta_2 x_1$$

and thus

$$\alpha_2/\beta_2 \; > \; x_1$$

This same type of analysis also shows us that
$dx_2/dt < 0$ when
$$\alpha_2/\beta_2 \; < \; x_1$$

There are two ways to understand results like this.
One is directly from the kind of algebraic analysis
just given. The other is by means of graphs which
indicate the qualitative dependence of the
variable's dynamics on the values of the modelled
variables and parameters. Such graphs are called
"phase portraits". In the present case, we have
the phase portrait shown in Figure 2.1.

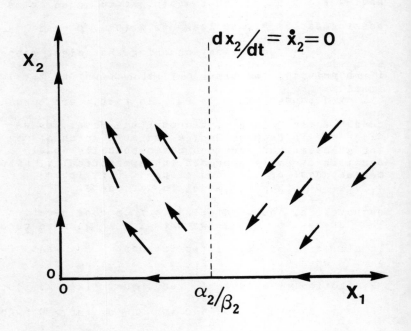

Figure 2.1. Phase portrait of
model (2.4) when $\alpha_1 \leq 0$ and $\alpha_2 > 0$.

Anyway, what does either algebra or phase portrait tell us about the population dynamics? If $x_1(0) > \alpha_2/\beta_2$, then either algebra or inspection

of the graph shows us that there must be some time τ , such that $x_1(t) < \alpha_2/\beta_2$ for all values of

$t > \tau$. But when $x_1 < \alpha_2/\beta_2$, $dx_2/dt > 0$.

Therefore we can conclude that $x_2(t)$ goes to ∞ as $t \to \infty$.

Finally, you should note that there was nothing special about casting species 1 for the unfortunate role of the species to go extinct in this local habitat. The whole analysis would have been exactly the same if all the subscripts had been reversed. Thus we have performed a general analysis of the consequences for a competitor having a negative growth rate in this model.

Having disposed of the alternative possibilities, henceforward we assume that all α_i and β_i parameters are strictly positive.

Returning to model (2.4), it is easy to show that there are only two cases for which both dx_i/dt equal zero:

$$x_1^* = 0 \quad \& \quad x_2^* = 0 \qquad (2.5.1)$$

and

$$x_1^* = \alpha_2/\beta_2 \quad \& \quad x_2^* = \alpha_1/\beta_1 \qquad (2.5.2)$$

The stability of either equilibrium depends on the dynamics of small perturbations near them. If we define

$$\varepsilon_1 = x_1 - x_1^* \quad \& \quad \varepsilon_2 = x_2 - x_2^* \qquad (2.6)$$

we have

$$
\begin{aligned}
d\varepsilon_i/dt &= dx_i/dt = x_i(\alpha_i - \beta_i x_j) \\
&= (\varepsilon_i + x_i^*)[\alpha_i - \beta_i(\varepsilon_j + x_j^*)] \\
&= \varepsilon_i[\alpha_i - \beta_i x_j^*] - \beta_i \varepsilon_i \varepsilon_j \\
&\quad + x_i^*[\alpha_i - \beta_i \varepsilon_j] - \beta_i x_i^* x_j^* \\
&= \varepsilon_i[\alpha_i - \beta_i x_j^*] + x_i^*[\alpha_i - \beta_i \varepsilon_j] \\
&\quad - \beta_i x_i^* x_j^* + O(\varepsilon_i^2)
\end{aligned}
$$

For small ε_i, we have

$$d\varepsilon_1/dt \approx \varepsilon_1[\alpha_1 - \beta_1 x_2^*] + x_1^*[\alpha_1 - \beta_1 \varepsilon_2]$$

$$- \beta_1 x_1{}^* x_2{}^*$$

$$d\epsilon_2/dt \approx \epsilon_2[\alpha_2 - \beta_2 x_1{}^*] + x_2{}^*[\alpha_2 - \beta_2 \epsilon_1] \qquad (2.7)$$

$$- \beta_2 x_1{}^* x_2{}^*$$

For equilibrium (2.5.1), (2.7) becomes

$$d\epsilon_1/dt \approx \alpha_1 \epsilon_1 \quad \& \quad d\epsilon_2/dt \approx \alpha_2 \epsilon_2$$

Since we are now assuming that $\alpha_i > 0$ for both i

we must have both $d\epsilon_i/dt > 0$, so that this

equilibrium is necessarily unstable. For
equilibrium (2.5.2), (2.7) becomes

$$d\epsilon_1/dt \approx \epsilon_1[\alpha_1 - \beta_1(\alpha_1/\beta_1)]$$

$$+ [\alpha_2/\beta_2][\alpha_1 - \beta_1 \epsilon_2]$$

$$- \beta_1[\alpha_1/\beta_1][\alpha_2/\beta_2]$$

$$= \alpha_1 \alpha_2/\beta_2 - \alpha_2 \beta_1 \epsilon_2/\beta_2$$

$$- \alpha_1 \alpha_2/\beta_2$$

$$= -\alpha_2 \beta_1 \epsilon_2/\beta_2$$

Now there are two ways to derive the analogous
expression for $d\epsilon_2/dt$. One is to follow the same

line of algebraic simplification as that just used
to find $d\epsilon_1/dt$. The other is to notice that the

original expressions for the $d\epsilon_i/dt$ were

symmetrical, with appropriate changes of subscript.
Therefore the final reduced form of (2.7) should
retain this same symmetry, since the equilibrium
(2.5.2) has exactly the same type of symmetry.
Thus we say, "by symmetry",

$$d\epsilon_2/dt \approx -\alpha_1 \beta_2 \epsilon_1/\beta_1$$

In matrix notation, we have

$$\begin{bmatrix} d\epsilon_1/dt \\ d\epsilon_2/dt \end{bmatrix} = \begin{bmatrix} 0 & -\alpha_2 \beta_1/\beta_2 \\ -\alpha_1 \beta_2/\beta_1 & 0 \end{bmatrix} \begin{bmatrix} \epsilon_1 \\ \epsilon_2 \end{bmatrix} \qquad (2.8)$$

or

$$\underline{\dot{\epsilon}} = J\underline{\epsilon} \qquad (2.9)$$

where, as in Section 1.4, we underline to indicate
a vector and use capitals to indicate a matrix.
The notational innovation in equation (2.9) is the
use of a "·" to indicate the first derivative with
respect to time; this is a common form of notation.
    Equations (2.8) and (2.9) represent what are
known as "linear systems with constant
coefficients". There are a great many mathematical
tools available for the analysis of such systems.
In particular, it is a well-known application of
matrix theory (e.g., Pullman, 1976, Chapter 3) that
the asymptotic behaviour of such linear systems
depends on the "eigenvalues" of matrix J, which
is called the "Jacobian matrix". The eigenvalues
of a matrix J are given by an equation which you
have already seen:
$$|J - \lambda I| = 0$$
Recall that the "$|\ \ |$" refers to the determinant
of the matrix between the parallel lines. [We
haven't yet had any formulae for calculating
determinants; we will write one down shortly.] The
eigenvalues are the values of λ for which this
equation is true, that is, the roots of this
equation. [In Section 1.4, the largest such λ was
the intrinsic rate of increase of an age-structured
population.] The dependence of the system's
behaviour on the eigenvalues is fairly simple for
continuous-time models: if the real part of each
and every eigenvalue is less than zero, all ε_i go

to 0 as $t \to \infty$, giving local asymptotic stability
of the equilibrium. [The real part is specified in
this rule because, in general, the λ's may be
complex numbers, as in the last sub-section of
Section 1.3.] The reason for this is that the
solution to equation (2.9) has the following form

$$\underline{\varepsilon}(t) = \underline{\varepsilon}(0)e^{Jt}$$
where the exponential function of the matrix is
analogous to the scalar,
$$\exp\{J\} = I + J + J^2/2! + J^3/3! + . .$$
As you should recall from Section 1.4, the
eigenvalues give the proportion by which a vector
is increased (or decreased) by matrix
multiplication once a great many of these
multiplications have taken place. Thus eigenvalues
are scalars which give the net effect of numerous
matrix multiplications. If the real parts of J's

eigenvalues are all less than zero, then J^n should

Competition

decrease faster than t^n increases, making $\exp\{Jt\}$ → 0 as $t \to \infty$. All this is the vector analogue of the requirements for stability characteristic of the models of Sections 1.1 and 1.2, in which negative slopes of the $r(x)$ functions were required for asymptotic stability.

If we let the j-th element in the i-th row of a two-by-two matrix M be represented by m_{ij}, then

$$| M | = m_{11}m_{22} - m_{12}m_{21} \qquad (2.10)$$

Thus the eigenvalues of J are the roots of

$$|J - \lambda I| = \begin{vmatrix} -\lambda & -\alpha_2\beta_1/\beta_2 \\ -\alpha_1\beta_2/\beta_1 & -\lambda \end{vmatrix}$$

$$= \lambda^2 - \alpha_1\alpha_2$$

which must be

$$\lambda_{\pm} = \pm\sqrt{\alpha_1\alpha_2} \qquad (2.11)$$

Since both α_i are greater than zero, both λ's are real, with $\lambda_+ > 0$ and $\lambda_- < 0$. Thus the stability criterion for equilibrium (2.5.2) is not met, since one of the eigenvalues has a real part greater than zero. Therefore model (2.4) never has asymptotically stable equilibria.

It is possible to solve system (2.4) directly to shed more light on what the population dynamics could possibly be getting up to. If we divide equation (2.4.2) by equation (2.4.1), we get

$$dx_2/dx_1 = x_2(\alpha_2-\beta_2x_1)/[x_1(\alpha_1-\beta_1x_2)] \qquad (2.12)$$

With this form, it is possible to separate the two variables, x_1 and x_2, and then integrate to find their trajectories as functions of time. First, we reorganize equation (2.12):

$$x_2^{-1}(\alpha_1-\beta_1x_2)dx_2 = x_1^{-1}(\alpha_2-\beta_2x_1)dx_1$$

or

$$[\alpha_1/x_2]dx_2 - \beta_1dx_2 = [\alpha_2/x_1]dx_1 - \beta_2dx_1$$

Now we integrate both sides, with the limits of integration being $x_i(0)$ and $x_i(t)$ for integration with respect to x_i:

56

giving
$$\alpha_1 \int x_2^{-1} dx_2 - \beta_1 \int dx_2 = \alpha_2 \int x_1^{-1} dx_1 - \beta_2 \int dx_1$$

$$\alpha_1 [\ln\{x_2(t)\} - \ln\{x_2(0)\}] - \beta_1 [x_2(t) - x_2(0)]$$

and thus
$$= \alpha_2 [\ln\{x_1(t)\} - \ln\{x_1(0)\}] - \beta_2 [x_1(t) - x_1(0)]$$

$$\alpha_1 \ln[x_2(t)/x_2(0)] - \beta_1 [x_2(t) - x_2(0)]$$

$$= \alpha_2 \ln[x_1(t)/x_1(0)] - \beta_2 [x_1(t) - x_1(0)] \qquad (2.13)$$

The unfortunate thing about equation (2.13) is that it gives the dynamics of one variable in terms of the other, instead of giving the dynamics of the two variables in terms of time. Thus equation (2.13) is not readily interpretable, by itself. However, you will see shortly the use which can be made of an equation of this kind.

Let us now turn to the global dynamics of the model. [The term "global" is in contrast to the "local" adjective used to indicate study of a model in the neighbourhood of equilibria. Evidently, global is to be taken to mean the dynamics "everywhere".] Obviously population dynamics models are only concerned with population densities x_i where $x_i \geq 0$, the "first quadrant" of real

two-dimensional space in the case of models with only two species. Let us subdivide this first quadrant into four sectors, as in Figure 2.2. From the equations of model (2.4), it is clear that $dx_1/dt > 0$ for $x_2 < \alpha_1/\beta_1$ while $dx_2/dt > 0$ for

$x_1 < \alpha_2/\beta_2$, and conversely. Thus there are four

different types of system "flow" (meaning trajectories of population density change) corresponding to the four sectors delimited by the lines $x_2 = \alpha_1/\beta_1$ and $x_1 = \alpha_2/\beta_2$, as indicated

by the directions of the arrows of Figure 2.2. In Sector I, both x_i decline. In Sector III, both

x_i increase. In Sectors II and IV, one of the

x_i's increases, while the other decreases. The

question at this stage is, do solution trajectories in sectors II and IV allow the extinction of one species with the continued growth of the other? Consider the LHS of equation (2.13). Let

57

$x_2(t) \to \infty$; it can then be shown that the limit of

the LHS is $-\infty$. [This is in fact to be done in the Exercises.] But when the LHS is approaching $-\infty$, we must have the RHS of (2.13) also approaching $-\infty$. For positive $x_1(t)$, this is

possible under only two circumstances. One is if $x_1(t)$ too goes to ∞ . But consultation of our

phase portrait, Figure 2.2, shows that this is not possible. Or, rather, if it occurs, perhaps as an initial condition, both x_1 and x_2 will quickly

fall. And, in any case, our primary interest is in the population dynamics in Sectors II and IV. Interestingly, the other case in which the RHS of (2.13) will approach $-\infty$ is when $x_1(t)$ approaches

0 , because then $\ln[x_1(t)/x_1(0)]$ approaches $\ln 0$

which in fact has $-\infty$ as its limit. Again, by symmetry, we can reverse this analysis for the case when $x_1(t) \to \infty$. So we can conclude that as $t \to \infty$

$x_j(t) \to 0$, for $i \neq j$. Thus we have the system

trajectories shown in Figure 2.2. It is apparent that both Sectors I and III must be left, while once the trajectory is in Sector II, $x_2(t)$

will increase without bound, while $x_1(t)$

will fall to zero. In Sector IV, $x_1(t)$ increases

without bound, while $x_2(t)$ falls toward zero.
There is one point which is of particular interest, that of equilibrium (2.5.2). The question is how the model's local dynamics near that equilibrium relate to the global dynamics. Consider the linearized system about this point, as given by equations (2.8) or (2.9), the eigenvalues of this system being given by equation (2.11). Associated with these eigenvalues are two eigenvectors, say $\underline{v}_+$ and $\underline{v}_-$, corresponding to

eigenvalues λ_+ and λ_- . Locally, the latter

eigenvector defines an axis along which trajectories flow toward the equilibrium, as shown in Figure 2.3.

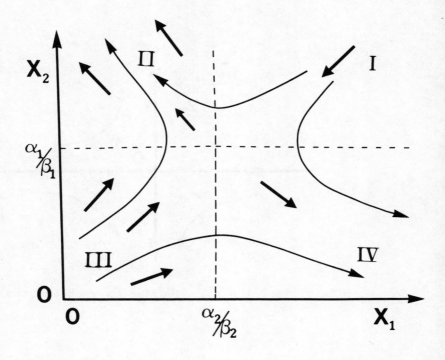

Figure 2.2. Trajectories of model
(2.4) which conform to the
constraints of equation (2.13).

Locally, the two eigenvectors provide bases for
system-state vectors, so that it would be possible,
in principle, to write an equation for the dynamics
of the model in terms of changes to the
eigenvector-components of a particular trajectory,
where this equation would have the following form:

$$\underline{\varepsilon}(t) \quad = \quad c_1 e^{\lambda_+ t} \underline{v}_+ \quad + \quad c_2 e^{\lambda_- t} \underline{v}_-$$

where the c_i are scalars chosen so that

$$\underline{\varepsilon}(0) \quad = \quad c_1 \underline{v}_+ \quad + \quad c_2 \underline{v}_-$$

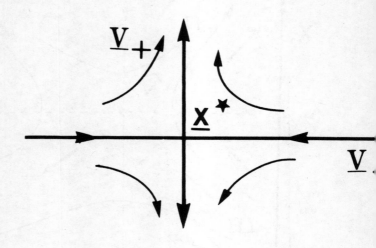

Figure 2.3. Local system dynamics
about the equilibrium given by
(2.5.2).

From this equation, it is clear that, as t goes
to ∞, $\underline{\epsilon}(t)$ becomes arbitrarily large, unless
$c_1 = 0$. When $c_1 = 0$, $\underline{\epsilon}(0) = c_2 \underline{v}_-$,

and $\underline{\epsilon}(t)$ converges on the origin [which

represents the equilibrium point] as t increases
without bound. From Figure 2.3, as well as the
vector algebra, it is evident that this approach to
the equilibrium proceeds only along the axis
defined by $\underline{v}_-$. On one side of this axis, the

trajectory diverges from the equilibrium in the
direction of Sector II, while on the other side it
diverges in the direction of Sector IV. Since
there is a unique line dividing these two

flow-patterns in the vicinity of the equilibrium, there must be a unique "separatrix" splitting the phase-portrait into two regions, trajectories originating in one region proceeding toward Sector II while trajectories originating in the other proceed toward Sector IV, as shown in Figure 2.4.

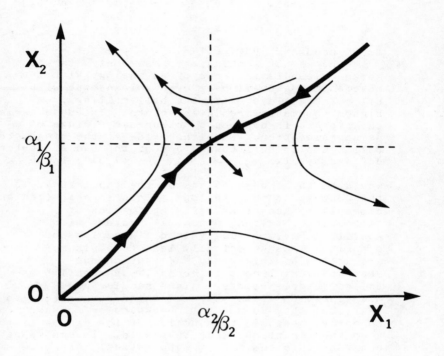

Figure 2.4. Separatrix defining the regions from which trajectories proceed toward one of Sectors II and IV.

In fact, since we know one point on the separatrix, that of the unstable equilibrium (2.5.2), and we have the general formula for all $\underline{x}(t)$ trajectories, given by (2.13), we can calculate the formula for the separatrix explicitly. The time variable in equation (2.13) is implicit, in that

(2.13) only gives curves in two-dimensional (x_1, x_2) space. Thus we can take equilibrium

(2.5.2) as the initial value of a trajectory and reverse the direction of time to find other values along the separatrix:

$$\alpha_1 \ln[x_2(t)/(\alpha_1/\beta_1)] - \beta_1[x_2(t) - \alpha_1/\beta_1] \quad (2.14)$$

$$= \alpha_2 \ln[x_1(t)/(\alpha_2/\beta_2)] - \beta_2[x_1(t) - \alpha_2/\beta_2]$$

or

$$\alpha_1 \ln[\beta_1 x_2/\alpha_1] - \beta_1 x_2 + \alpha_1 = \alpha_2 \ln[\beta_2 x_1/\alpha_2] - \beta_2 x_1 + \alpha_2$$

(cf. Freedman, 1980, p.145).

One way of looking at phase-portraits with separatrices is in terms of an analogy with the movement of a small ball over a landscape with hills and valleys, with the ball rolling from higher points toward lower points. In these terms, the separatrix defines a "razor-back" ridge of raised elevation. On either side of the ridge, the ball tends to roll toward one of the two attractors defined by $(x_1, x_2) = (0, \infty)$ or $(\infty, 0)$, which

would be the lowest points on the topography, in the context of this analogy. Thus separatrices are commonly spoken of as defining "basins of attraction". Planar dynamical systems with equilibria having real-valued eigenvalues of opposite signs generically lead to basins of attraction of this kind, and thus unique separatrices. For this reason, such equilibria are called "saddle-points", since saddles have mid-lines, on either side of which people tend to fall, but they do not fall backward or forward normally, because the function of the saddle is to keep them in place along that axis. Though we have not gone into their full mathematical depth, we shall assume henceforth that, in two-dimensional models - models of two-species interactions, such unique separatrices are generated by equilibria with real eigenvalues of opposing signs. See Hirsch and Smale (1974) for a fuller treatment of such dynamical system properties.

One Carrying Capacity
Suppose that one of the competitor species is subject to self-limitation in its population growth, while the other is not. In Lotka-Volterra models, such self-limitation is represented by logistic equation terms, giving the following system of ordinary differential equations:

$$dx_1/dt = \alpha_1 x_1 [1 - x_1/K - \beta_1 x_2/\alpha_1] \quad (2.15.1)$$

$$dx_2/dt = x_2 [\alpha_2 - \beta_2 x_1] \quad (2.15.2)$$

where species 1 has the strictly positive carrying capacity K , when growing on its own.

If we have both $\alpha_i < 0$, and $x_1(0) < K$,

then both dx_i/dt are less than zero for all t ,

so that $x_i(t)$ falls toward zero as $t \to \infty$.

If $\alpha_2 > 0$ while $\alpha_1 = 0$, we are back to one

of the "no carrying capacity" cases already analysed.

If $\alpha_1 > 0$ and $\alpha_2 = 0$, we have

$$dx_2/dt = -\beta_2 x_1 x_2 \quad (2.16.2)$$

in place of (2.15.2). If $x_1(0) > 0$ and

$x_2(0) > 0$, then $x_2(t)$ converges to zero. As it

does so, equation (2.15.1) approaches
$$dx_1/dt = \alpha_1 x_1 [1 - x_1/K]$$

which is simply the logistic equation. Thus, as t goes to ∞ we expect $x_1(t)$ to approach K

asymptotically. What does the phase-portrait look like, then? We already know that $dx_2/dt < 0$ for

all "interior" points (those with both $x_i > 0$), so

the remaining question is the pattern of dx_1/dt

values as a function of x_1 and x_2 . Obviously,

$dx_1/dt = 0$ for $x_1 = 0$, but this is of little

interest. If $x_1 > 0$ then $dx_1/dt = 0$ if and only if
$$1 - x_1/K - \beta_1 x_2/\alpha_1 = 0$$

which requires
$$x_2 = \alpha_1/\beta_1 - \alpha_1 x_1/[K\beta_1]$$
$$\quad (2.17)$$
$$= [1 - x_1/K]\alpha_1/\beta_1$$

Thus the values for which $dx_1/dt = 0$ define a

63

straight line in (x_1, x_2). Such a line is called a "zero-isocline", or simply an "isocline". The resulting phase portrait is shown in Figure 2.5.

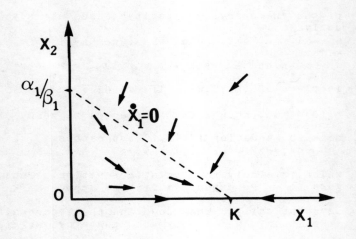

Figure 2.5. The phase portrait for model (2.15) when $\alpha_2 = 0$ and $\alpha_1 > 0$.

From (2.15.1), with $\alpha_2 = 0$, for x_2 below the isocline, $dx_1/dt > 0$, while for x_2 above the zero-isocline, $dx_1/dt < 0$. Similarly, for x_1 to the right of the isocline, $dx_1/dt < 0$, while for x_1 to the left of the isocline,

64

$dx_1/dt > 0$. Thus the system globally approaches the neighbourhood of $(x_1,x_2) = (K,0)$.

From (2.15.1), for x_2 below the zero-isocline, above the x_1 axis, interest centres on the local dynamics about this point. Again, we perform a perturbation analysis about this point, taking

$$x_1(t) = \varepsilon_1(t) + K$$

and

$$x_2(t) = \varepsilon_2(t)$$

both ε_i small in magnitude. As before, we derive a linearized model as an approximation to the dynamics about the equilibrium, finding

$$
\begin{aligned}
d\varepsilon_1/dt = dx_1/dt &= \alpha_1 x_1 [1 - x_1/K - \beta_1 x_2/\alpha_1] \\
&= \alpha_1(\varepsilon_1+K) [1 - (\varepsilon_1+K)/K - \beta_1\varepsilon_2/\alpha_1] \\
&= \alpha_1(\varepsilon_1+K) [-\varepsilon_1/K - \beta_1\varepsilon_2/\alpha_1] \\
&= -\alpha_1\varepsilon_1 - \beta_1 K\varepsilon_2 + O(\varepsilon_i^2) \\
&\approx -\alpha_1\varepsilon_1 - \beta_1 K\varepsilon_2 \qquad (2.18.1)
\end{aligned}
$$

and

$$
\begin{aligned}
d\varepsilon_2/dt = dx_2/dt &= -\beta_2 x_1 x_2 \\
&= -\beta_2(\varepsilon_1+K)\varepsilon_2 = -\beta_2 K\varepsilon_2 + O(\varepsilon_1\varepsilon_2) \\
&\approx -\beta_2 K\varepsilon_2 \qquad (2.18.2)
\end{aligned}
$$

In matrix notation, this gives us

$$
\begin{bmatrix} d\varepsilon_1/dt \\ d\varepsilon_2/dt \end{bmatrix}
=
\begin{bmatrix} -\alpha_1 & -\beta_1 K \\ 0 & -\beta_2 K \end{bmatrix}
\begin{bmatrix} \varepsilon_1 \\ \varepsilon_2 \end{bmatrix}
\qquad (2.19)
$$

This perturbation method has been our standard way of generating these linearized equations about equilibria, but in fact there is a less cumbersome way of going about the derivation of Jacobian matrices. The ij-th element of such matrices turns out to be the first partial derivative of the formula for dx_i/dt with respect to x_j .

65

[A partial derivative is just like a normal derivative with the "other" variables, those other than the one with respect to which the derivative is being taken, being taken as constants.] Here we will follow conventional typographical notation, and let $\partial f(x,y)/\partial y$ indicate the first partial derivative of $f(x,y)$ with respect to y. With this notation, the local dynamics of a perturbation to variable x_i about some equilibrium $\underline{x}^*$,

say ε_i, has the following form when the expression for dx_i/dt is given by $f_i(\underline{x})$

$$d\varepsilon_i/dt \approx \Sigma_j [\partial f_i(\underline{x})/\partial x_j |_{\underline{x}^*} \varepsilon_j] \qquad (2.20)$$

If we let J_{ij} represented the ij-th element of the Jacobian matrix, then we have

$$J_{ij} = \partial f_i(\underline{x})/\partial x_j \qquad (2.21)$$

It is to be checked in the Exercises that the application of these formulae to equilibrium $(K,0)$ readily recovers equation (2.19). It is also to be shown in the Exercises that the eigenvalues of the Jacobian at this equilibrium are $-\alpha_1$ and $-\beta_2 K$,

both less than zero on our assumptions. Thus, the equilibrium is locally asymptotically stable. This, together with our characterization of the global dynamics, shows us that the population dynamics result in global convergence to $(K,0)$ as t goes to ∞. This equilibrium therefore has the same property of global asymptotic stability that was exhibited by K in model (1.15). Applying this same method to the origin also shows that it is unstable, as is to be done in the Exercises.

Now let us consider the more general case with α_1, α_2, β_1, β_2, and K all strictly greater than zero. Again the curve for $dx_1/dt = 0$ is

given by equation (2.17). We also have a zero-isocline for dx_2/dt given by

$$x_1 = \alpha_2/\beta_2 > 0$$

which is perpendicular to the x_1 axis. There are only two qualitatively distinct phase-portraits for

the system when it has positive parameters, and
these are given in Figures 2.6 and 2.7.

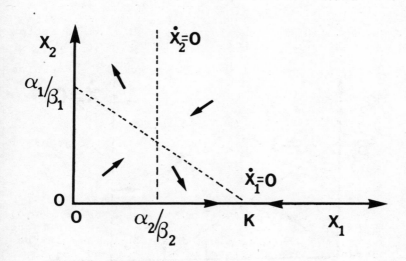

Figure 2.6. Phase-portrait of model
(2.15) with all parameters strictly
positive and $K > \alpha_2/\beta_2$.

There are always two equilibria:
$$\underline{x}_1{}^* = (0,0) \quad \& \quad \underline{x}_2{}^* = (K,0) \qquad (2.22)$$

In addition, if $K > \alpha_2/\beta_2$, the case shown in

Figure 2.6, there is a distinct third equilibrium
where the line for which $dx_1/dt = 0$ intersects the

zero-isocline of dx_2/dt . The location of this

point is readily obtained algebraically by
substituting $x_1 = \alpha_2/\beta_2$ into equation (2.17),
which gives

67

$$x_2 = \alpha_1 [1 - \alpha_2/\beta_2 K]/\beta_1$$

$$= \alpha_1/\beta_1 - \alpha_1 \alpha_2/\beta_1 \beta_2 K$$

and thus

$$\underline{x}_3^* = (\alpha_2/\beta_2, \alpha_1/\beta_1 - \alpha_1 \alpha_2/\beta_1 \beta_2 K) \qquad (2.23)$$

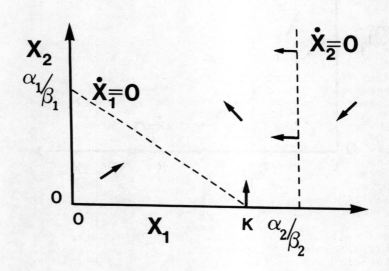

Figure 2.7. Phase-portrait of model (2.15) with all parameters strictly positive and $K \leq \alpha_2/\beta_2$.

The general form for the Jacobian matrix of the linearized system about any of these three equilibria is as follows, taking $\underline{x}_i^* = (x_{1,i}^*, x_{2,i}^*)$:

$$\left[\alpha_1 - 2\alpha_1 x_{1,i}^*/K - \beta_1 x_{2,i}^* \qquad -\beta_1 x_{1,i}^* \right] \qquad (2.24)$$

$$\left[\quad -\beta_2 x_{2,i}* \qquad\qquad \alpha_2 - \beta_2 x_{1,i}* \quad \right]$$

[This is to be shown in the Exercises.] It should be obvious to you that $\underline{x}_1*$ is unstable, but if it isn't you should use matrix (2.24) to convince yourself. Local asymptotic stability of $\underline{x}_2*$ requires $\beta_2 K > \alpha_2$, as is to be shown in the Exercises. You are also to show that, when $\underline{x}_3*$ exists, it is always unstable, with one negative and one positive eigenvalue. As for the case with no carrying capacities, $\underline{x}_3*$ is a saddle-point,

defining the end-point of a separatrix trajectory separating two basins of attraction. Thus the population dynamics give rise to explosive growth of species 2 with declining levels of species 1 in the case of the phase-portrait of Figure 2.6, while the dynamics in the case of Figure 2.7 depend on the initial conditions. Above and to the left of the separatrix, the dynamics are the same as for Figure 2.7. Below and to the right of the separatrix, the dynamics tend to eliminate species 2 and equilibration of the density of species 1 at its carrying capacity, equilibrium $\underline{x}_2*$. In general,

the self-limiting population has a greater risk of extinction due to competition, since it requires specific parametric conditions and specific initial conditions in order to prevail. Thus, in theory, self-limitation can endanger species survival, a result exactly contrary to the speculations of Wynne-Edwards (1962).

2.2 CLASSICAL LOTKA-VOLTERRA MODEL

All the Lotka-Volterra models discussed so far lack at least one of the minimum requirements for a plausible model of interspecific competition. The normal assumption in Lotka-Volterra theory is that there is density-dependent intraspecific and interspecific competition for all species. This necessarily gives rise to models where all species have carrying capacities as well as sensitivity to the competitive effects of the other species. In this framework, the minimum satisfactory model is as follows:

$$dx_1/dt = x_1(\alpha_1 - \alpha_1 x_1/K_1 - \beta_1 x_2) \qquad (2.25.1)$$

$$dx_2/dt = x_2(\alpha_2 - \alpha_2 x_2/K_2 - \beta_2 x_1) \qquad (2.25.2)$$

where all the parameters have the same interpretation as those of Section 2.2, and all are strictly positive. This is the Lotka-Volterra model of competition which many of you will have read about in introductory ecology textbooks.

There are three obvious equilibria:

$$\underline{x}_1* = (0,0) \qquad (2.26)$$

$$\underline{x}_2* = (K_1,0) \qquad (2.27)$$

$$\underline{x}_3* = (0,K_2) \qquad (2.28)$$

all three of which always arise. [The existence of these equilibria should be easy for you to see by now. If not, then review Section 2.1 again. If you have not already done the exercises associated with that section, and are having trouble with this section, then you should do the exercises now.]

In addition, there may be a fourth equilibrium with both x_1 and x_2 greater than zero. For this equilibrium to arise, equations (2.25.1) and (2.25.2) must simultaneously equal zero for non-zero values of both x_i. These equations are composed of two factors multiplied together: an x_i and an expression which reflects the impact of both species on the reproductive rate of individuals of species i. Since the former cannot be zero in this case, we seek conditions in which both of these second expressions are simultaneously zero, as follows.

$$\alpha_1 - \alpha_1 x_1/K_1 - \beta_1 x_2 = \alpha_2 - \alpha_2 x_2/K_2 - \beta_2 x_1 = 0 \qquad (2.29)$$

If we define

$$\Delta = \alpha_1 \alpha_2 - \beta_1 \beta_2 K_1 K_2$$

then you are expected to show in the Exercises that condition (2.29) is met when

$$\underline{x}_4* = (\alpha_2 K_1 [\alpha_1 - \beta_1 K_2]/\Delta, \alpha_1 K_2 [\alpha_2 - \beta_2 K_1]/\Delta) \qquad (2.30)$$

If the elements of $\underline{x}_4*$ derivable from this formula are negative or undefined, then the equilibrium is not admissible, meaning that it is not located in the first quadrant of the two-dimensional space defined by all values of the two x_i . [The first quadrant is that for which both $x_i > 0$.]

As before, the local dynamics about the equilibria can be understood in terms of a linearized system giving the first-order effects of initial perturbations from equilibrium upon further deviations from the equilibrium. The Jacobian matrix for these linearized systems has the following general form, as you are to show in the Exercises.

$$
\begin{bmatrix}
\alpha_1 - 2\alpha_1 x_{1,i}*/K_1 - \beta_1 x_{2,i}* & -\beta_1 x_{1,i}* \\
\\
-\beta_2 x_{2,i}* & \alpha_2 - \beta_2 x_{1,i}* - 2\alpha_2 x_{2,i}*/K_2
\end{bmatrix} \tag{2.31}
$$

where the equilibrium point is given by
$$\underline{x}_i* = (x_{1,i}*, x_{2,i}*)$$

as before.

You should find it readily apparent that equilibrium $\underline{x}_1*$ is unstable, since the

eigenvalues of matrix (2.31) when both $x_{i,j}*$ are

zero are α_1 and α_2 , which have both been assumed

to be positive. That these are the eigenvalues you should also find intuitively natural by this point, since eigenvalues give measures of the magnitudes of the system "flow" about an equilibrium point. The α_i are analogues of the intrinsic rate of

increase parameters from Chapter One, and thus they are naturally cast as the eigenvalues of the model when both population densities are low.

The instability of the origin shows us that the system dynamics must move outward along the axes or into the interior of the first quadrant. The system dynamics may then be sorted out with the aid of phase-plane plots. There are four major distinguishable cases, which are depicted in Figures 2.8 to 2.11. These four cases are algebraically defined by the relative magnitude of the two K_i with respect to the magnitude of the

α_j/β_j ratio, where i does not equal j .

Certain basic features characterize all these phase-portraits. Most importantly, below and to the left of the $dx_i/dt = 0$ isocline $dx_i/dt > 0$ while above and to the right of this isocline, $dx_i/dt < 0$. [Remember, $\dot{x}_i = dx_i/dt$, when comparing the text with the figures; the LHS notation is normally used in the figures for the sake of tidiness.] Thus, near the origin, the

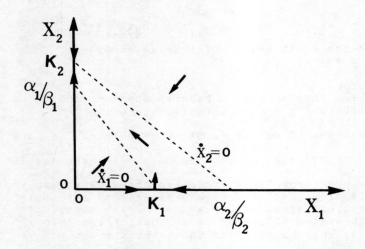

Figure 2.8. Isoclines and qualitative system flow for model (2.25) when $K_2 > \alpha_1/\beta_1$ and $K_1 \leq \alpha_2/\beta_2$.

system tends to move toward the upper right-hand corner, while in the upper right-hand corner the system tends to move toward the origin. This tells us that the system should eventually end up

somewhere near the isoclines which proceed
diagonally from the upper left-hand side to the
lower right-hand side of the quadrant.
    Consider the case shown in Figure 2.8, with
$K_2 > \alpha_1/\beta_1$ and $K_1 \leq \alpha_2/\beta_2$. Given the values of

the dx_i/dt on either side of their isoclines, we

must have $dx_1/dt < 0$ and $dx_2/dt > 0$ between the

isoclines. This will tend to drive the dynamics
toward the upper left-hand extreme of the domain
between the isoclines, near $(0,K_2)$, one of the

three admissible equilibria. The third admissible
equilibrium, $(K_1,0)$ can be readily shown to be

unstable, as is to be done in the Exercises. Thus
it is clear that the population dynamics will
approach some local neighbourhood of $(0,K_2)$, so

that its ultimate fate must depend on the local
dynamics about this point, which in turn depend on
the Jacobian matrix's eigenvalues there.
Irrespective of the magnitude of the value of the
system parameters, providing that they are
positive, the Jacobian at this point is readily
found from (2.31):

$$J(\underline{x}_3^*) = \begin{bmatrix} \alpha_1 - \beta_1 K_2 & 0 \\ \\ -\beta_2 K_2 & -\alpha_2 \end{bmatrix} \qquad (2.32)$$

This matrix evidently has eigenvalues $-\alpha_2$ and

$\alpha_1 - \beta_1 K_2$. Thus, for all phase-portraits, $\underline{x}_3^*$ is

locally asymptotically stable if and only if
$$K_2 > \alpha_1/\beta_1 \qquad (2.33)$$

We can see immediately that the phase-portrait of
Figure 2.8 guarantees $K_2 > \alpha_1/\beta_1$, and thus the

local asymptotic stability of $(0,K_2)$, given

convergence to it. Since the phase-portrait also
shows convergence to this local neighbourhood, we
can conclude that $(0,K_2)$ is the global attractor

for the population dynamics resulting from the

two-species competition dynamics represented by Figure 2.8.

An exactly parallel analysis shows that $(K_1, 0)$ is the global attractor in the case

displayed in Figure 2.9. Such analysis also shows that

$$K_1 > \alpha_2/\beta_2 \qquad\qquad (2.34)$$

is the condition for local asymptotic stability of equilibrium $\underline{x}_2*$. In fact, this whole analysis is

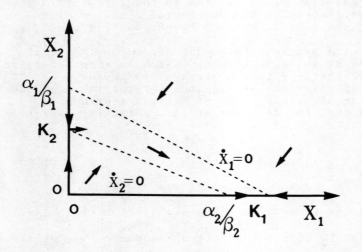

Figure 2.9. Isoclines and qualitative system flow for model (2.25) when $K_2 \leqq \alpha_1/\beta_1$ and

$K_1 > \alpha_2/\beta_2$.

superfluous, since there are no fundamental differences between species 1 and 2 in the competition model, so that they are only notationally distinct from the standpoint of algebra. Thus, one can immediately draw the parallel conclusions "by symmetry".

Turning to the cases of Figures 2.10 and 2.11, the local analyses of the dynamics about equilibria $\underline{x}_2*$ and $\underline{x}_3*$ which were just done also reveal that both of these equilibria are locally asymptotically stable in the case of Figure 2.11, while both are unstable in the case of Figure 2.10. However, $\underline{x}_4*$ is admissible in both these cases, so we need information about its stability properties for our analysis. From matrix (2.31), the Jacobian about this interior equilibrium is given by

$$
J(\underline{x}_4*) = \begin{bmatrix} -\alpha_1\alpha_2(\alpha_1-\beta_1 K_2)/\Delta & -\alpha_2\beta_1 K_1(\alpha_1-\beta_1 K_2)/\Delta \\ \\ -\alpha_1\beta_2 K_2(\alpha_2-\beta_2 K_1)/\Delta & -\alpha_1\alpha_2(\alpha_2-\beta_2 K_1)/\Delta \end{bmatrix}
$$

which we will designate matrix (2.35), where Δ is as before. In the case of Figure 2.10, when all boundary equilibria are unstable, we have $\alpha_j/\beta_j > K_i$, for i not equal to j. Thus the $\alpha_j - \beta_j K_i$ terms are then greater than zero. Moreover, we have

$$
\begin{aligned}
\Delta &= \alpha_1\alpha_2 - \beta_1\beta_2 K_1 K_2 \\
&> \alpha_1\alpha_2 - \beta_1\beta_2(\alpha_2/\beta_2)(\alpha_1/\beta_1) = 0
\end{aligned}
\tag{2.36}
$$

Thus all terms of matrix (2.35) are negative in this instance, since they all have negative numerators and positive denominators.

At this point, it is useful to consider some special properties of the eigenvalues of 2 x 2 matrices. Consider such a matrix, calling it M, with

$$
M = \begin{bmatrix} a & b \\ c & d \end{bmatrix}
$$

Then the eigenvalues of M are given by the roots of

$$\lambda^2 - \lambda(a+d) + ad - bc = 0$$

Explicitly, these roots are

$$\lambda_{\pm} = [a+d]/2 \pm \sqrt{(a+d)^2-4(ad-bc)}/2$$

Thus these roots have strictly negative real parts when

$$a + d < 0 \quad \& \quad ad - bc > 0 \qquad (2.37)$$

and there are no imaginary parts when

$$(a+d)^2 > 4(ad - bc) \qquad (2.38)$$

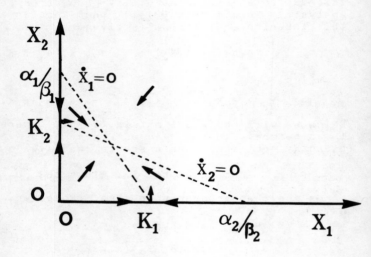

Figure 2.10. Isoclines and qualitative system flow for model (2.25) when $K_2 < \alpha_1/\beta_1$ and $K_1 < \alpha_2/\beta_2$.

In the case depicted in Figure 2.10, a and d are both negative so that the first of conditions (2.37) is met. For the second condition, we find that

$$ad - bc = \alpha_1^2 \alpha_2^2 (\alpha_2 - \beta_2 K_1)(\alpha_1 - \beta_1 K_2)/\Delta^2$$

$$- \alpha_1 \alpha_2 \beta_1 \beta_2 K_1 K_2 (\alpha_1 - \beta_1 K_2)(\alpha_2 - \beta_2 K_1)/\Delta^2$$

$$= \alpha_1 \alpha_2 (\alpha_1 - \beta_1 K_2)(\alpha_2 - \beta_2 K_1)\Delta^{-1} \qquad (2.39)$$

$$> 0$$

Thus condition (2.37) is met, guaranteeing local asymptotic stability of equilibrium $\underline{x}_4{}^*$ when the conditions for the phase-portrait of Figure 2.10 are met. From (2.38), there is no oscillation about the equilibrium during convergence if

$$a^2 + 2ad + d^2 > 4ad - 4bc$$

or

$$a^2 - 2ad + d^2 > -4bc$$

or

$$(a-d)^2 > -4bc$$

But $(a-d)^2 \geq 0$, while b and c are both negative, making the RHS of the inequality negative, so this condition too is met. Thus there is direct convergence to the interior equilibrium of Figure 2.10. From our discussion of the phase-portrait of Figure 2.9, you should be able to see that the system flow will bring the population densities to the region where the isoclines are. Between the two isoclines above and to the left of $\underline{x}_4{}^*$, $dx_2/dt < 0$ and $dx_1/dt > 0$, so that the system flow will bring the population densities down and to the right. Conversely, between the two isoclines below and to the right of $\underline{x}_4{}^*$,

$dx_1/dt < 0$ and $dx_2/dt > 0$, so that the system flow will bring the population densities up and to the left. Thus trajectories which start some distance away from $\underline{x}_4{}^*$ bring the population densities toward the neighbourhood of this equilibrium, thereby ensuring global convergence to it from all interior points.

In the case portrayed in Figure 2.11, the analysis of the interior dynamics is exactly reversed. In particular, while the a and d elements of the Jacobian matrix for equilibrium $\underline{x}_4{}^*$ remain negative, the sign of $ad - bc$ is reversed, as you are to show in the Exercises, which ensures that the interior equilibrium is unstable. Since both $(K_1,0)$ and

$(0,K_2)$ are locally asymptotically stable, we can infer that the interior equilibrium must be a saddle-point on a separatrix defining the boundary separating the basins of attraction of the two equilibria on the axes. You should now also have enough familiarity with phase-plane dynamical analysis to construct arguments showing that the

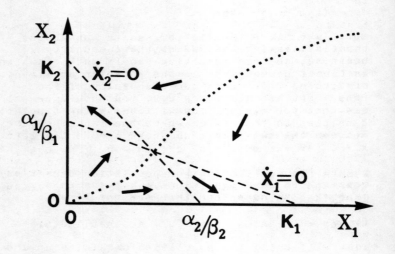

Figure 2.11. Isoclines and qualitative system flow for model (2.25) when $K_2 > \alpha_1/\beta_1$ and $K_1 > \alpha_2/\beta_2$.

system flow must bring all trajectories not beginning at the origin to some local neighbourhood of these two equilibria given the parameter values defining Figure 2.11.

The interpretation of these results is fairly straightforward. The larger the carrying capacity and the initial rate of increase, or the smaller the sensitivity to interspecific competition, of a species, then the more likely it is to prevail in the context of competition. If the carrying capacities of two competing species tend to be low, while they are not overly sensitive to interspecific competition, which will tend to be the case if two species have very different limiting resources, then the outcome of competition is more likely to be coexistence. If the carrying capacities of the two competitors tend to be high, while the sensitivity to interspecific competition is also high, then competition is likely to result in predominance of one species or the other, depending on the initial conditions. This would tend to be the case when the two species are more similar in their ecological requirements. All of these findings accord with intuition, although their scientific significance really depends on whether or not they are borne out in less simplified models, since it is generally not a good idea to found one's understanding entirely on the results of the simplest conceivable models.

2.3 GENERAL CONTINUOUS-TIME MODELS

Just as we did for single-species models of population growth, we will now consider models of competition in which the density-dependent net reproductive rate functions are not linear. Again, as we found for nonlinear population growth models, it is possible to infer a great deal about the population dynamics from qualitative graphical analysis of the system's dynamical properties. In so doing, we will be able to avoid monstrous algebraic expressions.

Following Rescigno and Richardson (1967), Hirsch and Smale (1974, pp.265-273) and Freedman (1980, pp.166-167), let us consider the following model:

$$dx_1/dt = x_1 f_1(x_1, x_2) \qquad (2.47.1)$$

$$dx_2/dt = x_2 f_2(x_1, x_2) \qquad (2.47.2)$$

In order that system (2.47) properly represents the population dynamics of competition, we require, for all admissible values of (x_1, x_2) ,

$$\partial f_1(x_1,x_2)/\partial x_2 \; < \; 0 \quad (2.48.1)$$

$$\partial f_2(x_1,x_2)/\partial x_1 \; < \; 0 \quad (2.48.2)$$

In order to forestall cases where either population density falls toward extinction, we will take

$$f_1(0,0) \; > \; 0 \quad \& \quad f_2(0,0) \; > \; 0 \quad\quad (2.49)$$

The effect of this assumption is to require that the origin be unstable. Furthermore, we assume that neither population can become indefinitely large when growing on its own. That is, we assume that there are K_1 and K_2 such that

$$f_1(K_1,0) = 0 \quad \& \quad f_2(0,K_2) = 0 \quad\quad (2.50)$$

where both K_i are strictly positive and

$f_1(x_1,0) < 0$ for $x_1 > K_1$, and similarly for f_2 .

Conversely, we take $f_1(x_1,0) > 0$ for $K_1 > x_1 > 0$

and likewise for f_1 . The effect of these

assumptions is to preclude multiple stable equilibria for either competitor when it is growing on its own. We also assume that there is a $K > 0$ such that both $f_i(x_1,x_2) < 0$ for x_1 or x_2

greater than or equal to K . The effect of this assumption is to preclude unbounded population growth of either competitor. Finally, we assume that, for $x_1 < K_1$, there is a unique, continuous,

and differentiable curve specified by the set

$$g_1 \; = \; \{ \; (x_1,x_2) \mid f_1(x_1,x_2) = 0 \; \} \quad\quad (2.51.1)$$

and similarly there is such a

$$g_2 \; = \; \{ \; (x_1,x_2) \mid f_2(x_1,x_2) = 0 \; \} \quad\quad (2.51.2)$$

These equations (2.51) give the system isoclines, all interior equilibria arising where the two isoclines intersect.

For almost all f_i's, g_1 and g_2 will

intersect at single points, though they may do so several times. Any mathematical specification of the f_i which does <u>not</u> have this property can be

approximated arbitrarily closely by a choice of the f_i which does; thus such cases are of no scientific interest. If we consider arbitrary g_i's meeting this condition, it is apparent that

they must divide the quadrant into regions, as shown in Figure 2.12. In the example shown, there are five such regions, R_1 to R_5. In general,

there will be at least three such regions. There will be an R_1 region, for which both $f_i > 0$.

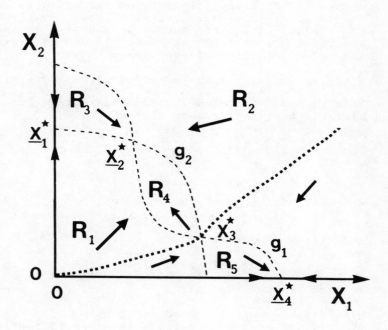

Figure 2.12. Graphical example of a phase-plane associated with system (2.47).

This follows from assuming that both $f_i(0,0) > 0$.

Likewise, there will always be a region like R_2

in Figure 2.12, for which both $f_i < 0$, from the
assumption concerning K above. All other regions,
like regions R_3 to R_5 in the figure, are
between R_1 and R_2 . Hirsch and Smale (1974,
pp.268-270) have shown that there are no limit
cycles within such regions. Any trajectory which
starts at the boundary of such a region either
flows out of it, never to return, or flows into it,
coming to rest at one of its boundary points. Thus
system (2.47) always flows to one of a finite
number of locally asymptotically stable equilibria.
    System equilibria may be on the axes or inside
the first quadrant of the phase-plane. Consider
the subset of internal equilibria. Let $f_{i,j} =$
$x_i * \partial f_i / \partial x_j$, for those $x_i *$ for which $f_i(\underline{x}_i *) = 0$.
Then we have the following Jacobian matrix for
linearized systems about $\underline{x}*$ values:

$$\begin{bmatrix} f_{1,1} & f_{1,2} \\ f_{2,1} & f_{2,2} \end{bmatrix}$$

with eigenvalues given by the roots of

$$(f_{1,1}-\lambda)(f_{2,2}-\lambda) - f_{1,2}f_{2,1} = 0 \qquad (2.52)$$

or

$$2\lambda_{\pm} = f_{1,1} + f_{2,2} \pm \sqrt{(f_{1,1}+f_{2,2})^2 + 4(f_{2,1}f_{1,2} - f_{1,1}f_{2,2})}$$

We have assumed that both $f_{1,2}$ and $f_{2,1}$ are
negative, so that the product of these two
derivatives must be positive. In order to forestall
$\lambda_+ > 0$, which we must if the equilibrium is to be
locally asymptotically stable, we must have

$$4(f_{2,1}f_{1,2} - f_{1,1}f_{2,2}) < 0 \qquad (2.53)$$

A necessary condition for this to be true is
$f_{1,1}f_{2,2} < 0$. This in turn requires that the $f_{i,i}$
be of the same sign. But if they are of the same
sign, then they must both be negative, since
otherwise one of the two eigenvalues must be
positive. The necessary and sufficient condition

for this inequality to arise is

or $\quad f_{1,1}f_{2,2} > f_{1,2}f_{2,1}$

$\quad -f_{2,1}/f_{2,2} > -f_{1,1}/f_{1,2}$

Interestingly, $-f_{i,1}/f_{i,2}$ gives the slope of g_i near the equilibrium. So that this condition is equivalent to the requirement that the slope of g_2 be greater than that of g_1 at the equilibrium. We have also shown that both $f_{i,i}$ must be negative at this equilibrium, which will be true if and only if the slopes of the g_i are negative. Thus our local asymptotic stability criteria can be readily ascertained by graphical inspection. If we have the type of isocline-isocline intersection depicted in Figure 2.13, then the equilibrium is locally asymptotically stable, and otherwise it isn't. This result is in fact illustrated by the stability properties of the $\underline{x}_4{}^*$ equilibrium of the classical Lotka-Volterra model of competition.

Now we consider axis-isocline intersections which constitute equilibria. We therefore must have an i such that $x_i{}^* = 0$ and $x_j{}^* = K_j$, for i not equal to j . The jj element of the Jacobian matrix still equals $f_{j,j}$, because $f_j = 0$. But now

$$\partial[dx_i/dt]/\partial x_j = x_i{}^*\partial f_i/\partial x_j = 0 \qquad (2.54)$$

giving a zero-value to all $f_{i,j}f_{j,i}$ terms in the equations for the eigenvalues. From equation (2.52), then, the eigenvalues are given by the two $f_{i,i}$. At boundary equilibria, we must have $f_{j,j}$ less than zero, because of our assumptions about the sign of f_j on either side of K_j . From the definition of the Jacobian we can readily find that the ii element is

$$f_{i,i} = f_i(x_i{}^*,x_j{}^*) = f_i(0,K_j) \qquad (2.55)$$

Therefore, if $f_i(0,K_j) < 0$ then the edge equilibrium is locally asymptotically stable, and otherwise it is not stable. This condition is

83

readily ascertained by visual inspection of the positions of the isoclines along the x_i axis. If the value of x_j at which the g_i isocline meets the x_i axis is above K_j , then the equilibrium is unstable, while it is locally stable otherwise. All told, with these rules, it is possible to assess the stability of all competition model equilibria by graphical inspection of system isocline plots.

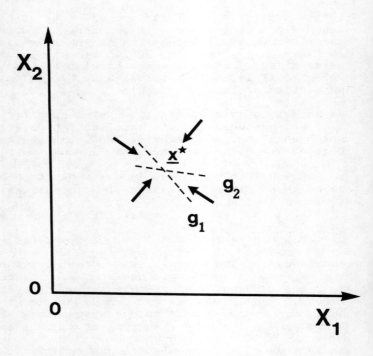

Figure 2.13. Isocline-isocline intersection configuration characteristic of locally asymptotically stable interior equilibria in general continuous-time models of competition.

There are two examples to hand. First is the classical Lotka-Volterra model, which meets all the assumptions built into the model of this section. Thus we have just rendered the algebraic work that we did on that model superfluous. Second is the case having the isoclines depicted in Figure 2.12. From our work to this point, we can readily see that equilibrium $\underline{x}_1*$ is unstable, $\underline{x}_2*$ is locally asymptotically stable, $\underline{x}_3*$ is a saddle-point, and $\underline{x}_4*$ is a second locally asymptotically stable equilibrium. There must be a separatrix passing through $\underline{x}_3*$ which defines the boundaries between the two local asymptotically stable equilibria. Thus, even without algebraic specification, the dynamical system giving rise to the configuration of isoclines can be understood to a fair degree, at least in terms of its eventual dynamics.

2.4 DISCRETE-TIME MODELS

As before, we turn from continuous-time models to discrete-time models. Rather than beginning with a discussion of a specific discrete-time model, let us proceed to discuss more general models of the same type as those treated in the sub-section just finished.

General Two-Species Models
We will preserve all our previous assumptions concerning carrying-capacities, existence of isoclines, and number of equilibria. These are features of the system's dynamical specification which do not depend on the assumption of discrete-time. Preserving them will enable us to understand better the effects of the change in the time-structure assumption upon the model's dynamical behaviour. However, you should note that now the isocline requirement is for values of (x_1, x_2) at which the $f_i[x_1, x_2] = 1$, given system equations having the following form, like that before:

$$x_1(t+1) = x_1(t)f_1[x_1(t), x_2(t)] \quad (2.56.1)$$

$$x_2(t+1) = x_2(t)f_2[x_1(t), x_2(t)] \quad (2.56.2)$$

If we consider an arbitrary interior equilibrium point, say $\underline{x}^*$, at which both f_i are equal to 1 , and we let $x_i(t) = x_i^* + \varepsilon_i(t)$, then we have, for example for $i = 1$,

$$\varepsilon_1(t+1) = x_1(t+1) - x_1^*$$

$$= x_1 f_1[x_1(t), x_2(t)] - x_1^*$$

$$= (x_1^* + \varepsilon_1) f_1[x_1^* + \varepsilon_1, x_2^* + \varepsilon_2] - x_1^*$$

Now we use a Taylor expansion to approximate $f_1[x_1, x_2]$:

$$f_1[x_1^* + \varepsilon_1, x_2^* + \varepsilon_2] = f_1[x_1^*, x_2^*] + \varepsilon_1 \partial f_1[x_1^*, x_2^*]/\partial x_1$$

$$+ \varepsilon_2 \partial f_1[x_1^*, x_2^*]/\partial x_2 + O(\varepsilon_i^2)$$

$$\approx 1 + \varepsilon_1 \partial f_1[x_1^*, x_2^*]/\partial x_1$$

$$+ \varepsilon_2 \partial f_1[x_1^*, x_2^*]/\partial x_2$$

Thus, for small ε_i , we take

$$\varepsilon_1(t+1) \approx \varepsilon_1 + \varepsilon_1 x_1^* f_{1,1} + \varepsilon_2 x_1^* f_{1,2} + O(\varepsilon_i^2)$$

with a parallel expression for $\varepsilon_2(t+1)$. The Jacobian matrix is then

$$\begin{bmatrix} 1 + f_{1,1} & f_{1,2} \\ f_{2,1} & 1 + f_{2,2} \end{bmatrix} \tag{2.58}$$

For local asymptotic stability, we must have $\underline{\varepsilon}(t)$ approaching the origin as time proceeds to infinity. Analogously to the requirements for asymptotic stability in discrete-time models of single-species population growth, this in turn requires that the eigenvalues of matrix (2.58) be strictly less than one in absolute magnitude. [Return to our discussion of the dynamics of age-structured population growth, as given by the Leslie matrix, if this is obscure to you.] Solving for these eigenvalues explicitly, we see that they are given by

$$2\lambda_{\pm} = 1 + f_{1,1} + f_{2,2} \tag{2.59}$$

Competition

$$\pm \sqrt{(f_{1,1}+f_{2,2})^2-4(f_{1,1}f_{2,2}-f_{1,2}f_{2,1})}$$

Using the same sort of arguments as those of the continuous-time case, it becomes apparent that (2.59) implies that local asymptotic stability requires <u>both</u> $f_{1,1}$ and $f_{2,2}$ to be negative,

since we must have $f_{1,1} + f_{2,2} < 0$ <u>and</u>

$$-4(f_{1,1}f_{2,2}-f_{1,2}f_{2,1}) < 0$$

simultaneously, even though $f_{1,2}f_{2,1} > 0$. Moreover, we again require

$$-f_{1,1}/f_{1,2} < -f_{2,1}/f_{2,2}$$

as in the continuous time case. Now, however, conformity with these inequalities does <u>not</u> guarantee asymptotic stability, if the negative factor $f_{1,1} + f_{2,2}$ is of sufficiently large

magnitude that the absolute magnitude of λ_- is greater than 1 . Thus we see that, once again, proceeding from continuous-time to discrete-time models may make local asymptotic stability of equilibria less likely. This is discussed further by Maynard Smith (1974, pp.67-68) and May (1974, pp.26-30).

The Hassell-Comins Model
There is little more to be said about the general discrete-time model of two competing species. To proceed farther, we need to consider a more concretely specified model. Among the few such studies of discrete-time competition theory is that of Hassell and Comins (1976). Using our previous notation, they examined the behaviour of a model which can be written as follows:

$$x_1(t+1) = \lambda_1 x_1(t)[1+a_1\{x_1(t)+\alpha x_2(t)\}]^{-b_1} \quad (2.60.1)$$

$$x_2(t+1) = \lambda_2 x_2(t)[1+a_2\{x_2(t)+\beta x_1(t)\}]^{-b_2} \quad (2.60.2)$$

with the $\lambda_i > 0$, the $a_i > 0$, and the $b_i > 0$.

For this to be a competition model, we must also have α and β strictly positive. In order to analyse this model, we first take $\theta_i = \lambda_i^{-1/b_i}$

and $\gamma_i = \theta_i a_i$, in order to rewrite model (2.60)

87

in the form

$$x_1(t+1) = x_1(t)[\theta_1 + \gamma_1\{x_1(t) + \alpha x_2(t)\}]^{-b}1 \quad (2.61.1)$$

$$x_2(t+1) = x_2(t)[\theta_2 + \gamma_2\{x_2(t) + \beta x_1(t)\}]^{-b}2 \quad (2.61.2)$$

At interior equilibria, we must have

and
$$[\theta_1 + \gamma_1\{x_1(t) + \alpha x_2(t)\}]^{-b}1 = 1$$

$$[\theta_2 + \gamma_2\{x_2(t) + \beta x_1(t)\}]^{-b}2 = 1$$

which gives rise to the linear isoclines for the system, as given by

$$x_1(t) + \alpha x_2(t) = [1-\theta_1]/\gamma_1 \quad (2.62.1)$$

$$x_2(t) + \beta x_1(t) = [1-\theta_2]/\gamma_2 \quad (2.62.2)$$

Thus, in some respects, the Hassell-Comins model is a discrete-time analogue of the continuous-time Lotka-Volterra competition model, in that both have linear isoclines.

As a direct result, the Hassell-Comins model also gives rise to a unique interior equilibrium which, when it is admissible, is given by

$$E_4 = (x_1^*, x_2^*)$$
$$(2.63)$$
$$= ([\alpha\mu_2 - \mu_1]/[1-\beta\alpha], [\beta\mu_1 - \mu_2]/[1-\beta\alpha])$$

where $\mu_i = \theta_i/\gamma_i = \theta_i/\theta_i a_i = a_i^{-1}$. There

are also two edge equilibria:

$$E_2 = ([1-\theta_1]/\gamma_1, 0) \quad \& \quad E_3 = (0, [1-\theta_2]/\gamma_2)$$

As usual, there is also the origin, $E_1 = (0,0)$.

For any of these equilibria, the Jacobian for the local linearized dynamics may be found from the following general expression

$$J(E) = \begin{bmatrix} 1 - \eta_1 & -\alpha\eta_1 \\ -\beta\eta_2 & 1 - \eta_2 \end{bmatrix} \quad (2.64)$$

where $\eta_i = x_i^* b_i \gamma_i$, as is to be shown in the

Exercises.

This Jacobian in turn gives rise to the characteristic polynomial

$$\lambda^2 + \lambda(\eta_1 + \eta_2 - 2) + \eta_1\eta_2(1-\alpha\beta) - (\eta_1+\eta_2-1) = 0 \quad (2.65)$$

In general, for a discrete-time system with characteristic equation of form

$$\lambda^2 + a_1\lambda + a_0 = 0 \quad\quad (2.66)$$

local asymptotic stability, due to $|\lambda_+| < 1$, is assured when

$$|a_0| < 1 \quad \& \quad |a_1| < 1+a_0 \quad\quad (2.67)$$

[This may be readily checked by examining the roots of equation (2.66) when $a_0 = 1$ and $a_1 = 1+a_0$.]

In the particular case of characteristic equation (2.65), conditions (2.67) do not assume particularly meaningful forms.
    Examination of the boundary dynamics shows that the competitors coexist only if

$$[1-\theta_1]/\gamma_1 > \alpha[1-\theta_2]/\gamma_2 \quad (2.68.1)$$

and

$$[1-\theta_2]/\gamma_2 > \beta[1-\theta_1]/\gamma_1 \quad (2.68.2)$$

as is to be demonstrated in the Exercises. This result is similar to that for the continuous-time Lotka-Volterra system, in which the isocline configuration of Figure 2.10, which required

$$\alpha_1/\beta_1 > K_2$$

and

$$\alpha_2/\beta_2 > K_1$$

sufficed to ensure coexistence of the two competitors. But unlike the Lotka-Volterra model, such coexistence in the discrete-time model does not imply convergence to E_4 ; conditions (2.68) can be met while conditions (2.67) are not. When such cases arise, the system exhibits limit cycles or chaotic behaviour, along the lines of that discussed in Chapter One's treatment of the discrete-time logistic model. Hassell and Comins (1976) provide numerical solutions exhibiting such behaviour, as well as a diagram giving critical parameter values for its occurrence.

2.5 SYMBIOSIS

Symbiosis is normally conceived in terms of the mutual facilitation of two species as a result of a considerable period of coevolution. It is doubtful that there are many species which conform to such an exacting conception. On the other hand, less direct facilitation of the net reproduction of one population by another may be quite common (May, 1981, p.95). Accordingly, the mathematical theory associated with symbiosis deserves some consideration.

Lotka-Volterra Models

Following Freedman (1980, pp.172-177), we begin by considering two-species models in which reproductive rates are given by linear functions of the populations' densities:

$$dx_1/dt = x_1[\alpha_1 - \alpha_1 x_1/K_1 + \beta_1 x_2] \quad (2.69.1)$$

$$dx_2/dt = x_2[\alpha_2 - \alpha_2 x_2/K_2 + \beta_2 x_1] \quad (2.69.2)$$

where all parameters are strictly positive, with interpretations like those before, except that now the β_i parameters measure the extent of mutual

facilitation, rather than interference or competition.

As for the Lotka-Volterra model of competition, there are two linear isoclines in the first quadrant, an interior equilibrium when these isoclines intersect, and three boundary equilibria. All system properties are readily derivable from the analysis of Section 2.2, with the sign of the β_i factors reversed. The two isoclines are as

follows:
$dx_1/dt = 0$ when
$$x_2 = [\alpha_1 x_1 - \alpha_1 K_1]/\beta_1 K_1 \quad (2.70.1)$$

$dx_2/dt = 0$ when
$$x_2 = [K_2\beta_2 x_1 + \alpha_2 K_2]/\alpha_2 \quad (2.70.2)$$

The three boundary equilibria are, as before, $(0,0)$, $(K_1,0)$, and $(0,K_2)$. The fourth

is given by
$$\underline{x}_4{}^* = (\alpha_2 K_1[\alpha_1 + \beta_1 K_2]\Delta^{-1}, \alpha_1 K_2[\alpha_2 + \beta_2 K_1]\Delta^{-1}) \quad (2.71)$$

where $\Delta = \alpha_1\alpha_2 - \beta_1\beta_2 K_1 K_2$, as before. Evidently,

if $\Delta \leq 0$ then (2.71) does not give an equilibrium in the first quadrant.

The Jacobian matrix for system equilibria is readily obtained as

$$\begin{bmatrix} \alpha_1 - 2\alpha_1 x_{1,i}*/K_1 + \beta_1 x_{2,i}* & \beta_1 x_{1,i}* \\ \\ \beta_2 x_{2,i}* & \alpha_2 - 2\alpha_2 x_{2,i}*/K_2 + \beta_2 x_{1,i}* \end{bmatrix} \quad (2.72)$$

From direct inspection of this matrix, it is apparent that the origin is unstable. At $(K_1, 0)$, the Jacobian matrix takes on the form

$$\begin{bmatrix} -\alpha_1 & \beta_1 K_1 \\ \\ 0 & \alpha_2 + \beta_2 K_1 \end{bmatrix}$$

with eigenvalues $-\alpha_1$ and $\alpha_2 + \beta_2 K_1 > 0$, so that

this equilibrium is always unstable. By symmetry, $(0, K_2)$ must also always be unstable. At the

interior equilibrium, the Jacobian matrix becomes

$$\begin{bmatrix} -\alpha_1 x_1*/K_1 & \beta_1 x_1* \\ \\ \beta_2 x_2* & -\alpha_2 x_2*/K_2 \end{bmatrix}$$

as is to be shown in the Exercises. This matrix has eigenvalues

$$\lambda_{\underline{+}} = -[\alpha_1 x_1*/K_1 + \alpha_2 x_2*/K_2]/2 \quad (2.73)$$
$$\underline{+} \sqrt{\{[\alpha_1 x_1*/K_1 + \alpha_2 x_2*/K_2]^2 - 4\Delta x_1* x_2*/K_1 K_2\}/2}$$

We have all parameters and state variables greater than zero, as well as $\Delta > 0$, so the real parts of the eigenvalues must be less than zero. Therefore, we can conclude that the interior equilibrium, when it exists, is locally asymptotically stable.

If we turn to the isocline phase-plane plots, we see that $\Delta \le 0$ gives rise to explosive growth of both populations, while $\Delta > 0$ gives a global asymptotically stable equilibrium specified by equation (2.71), as shown in Figures 2.14 and 2.15. However, the basic features of the system flow shown in Figure 2.14 are evidently implausible, in that both populations are growing to arbitrarily large sizes, in spite of normally being subject to intraspecific density-dependent population regulation. The critical parameter in determining which of the two situations arises is Δ. The requirement for the dynamics of Figure 2.14 is $\Delta \leq 0$, which in turn requires that the β_i

be sufficiently large. In words, this means that the symbiotic facilitation must be stronger than intraspecific population regulation. It is surely more reasonable to suppose that it is not, especially for indirect forms of mutual facilitation, so that we will have $\Delta > 0$, and thus the dynamics of Figure 2.15.

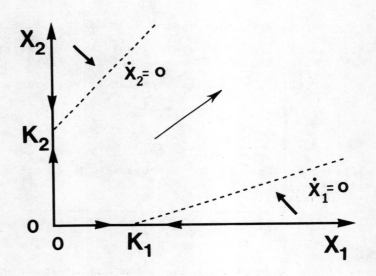

Figure 2.14. Phase-plane portrait of the population dynamics of model (2.69) with $\Delta \leq 0$.

General Continuous-Time Models
In the same spirit as our previous discussions of generalized models, we turn to less explicit models

of the form:

$$dx_1/dt = x_1 f_1(x_1,x_2) \qquad (2.74.1)$$

$$dx_2/dt = x_2 f_2(x_1,x_2) \qquad (2.74.2)$$

with $f_1/x_2 > 0$ and $f_2/x_1 > 0$, in order to represent the assumption of symbiosis. We assume: (i) $f_i(0,0) > 0$ for $i = 1$ & 2 ; (ii) the existence of two K_i such that $f_1(K_1,0) = 0$ and $f_2(0,K_2) = 0$ uniquely along the x_1 and x_2 axes;

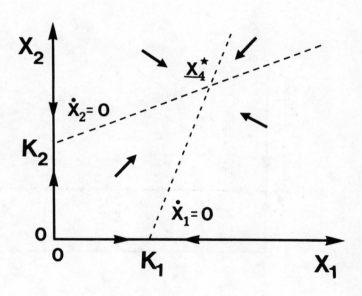

Figure 2.15. Phase-plane portrait of the population dynamics of model (2.69) with $\Delta > 0$.

as well as (iii) isoclines given by single curves for each dx_i/dt . Under these circumstances we

can draw general isocline phase-plane plots of two basic types. The first, shown in Figure 2.16, lacks interior equilibria, because the isoclines do not intersect in the first quadrant. The system flow brings the system state into the region between the two isoclines, and within this region both population densities increase without bound. The second type of phase-plane has interior equilibria, due to isocline-isocline intersections, as shown in Figure 2.17. These two cases are of course the analogues of those shown in Figures 2.14 and 2.15, respectively.

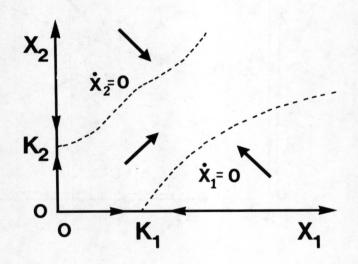

Figure 2.16. Phase-plane plot for model (2.74) when there are no intersections of the isoclines along which the $dx_i/dt = 0$.

Interior equilibria in the case of Figure 2.17 will have Jacobian matrices of the same form as those characteristic of model (2.47) from Section 2.3:

$$\begin{bmatrix} f_{1,1} & f_{1,2} \\ f_{2,1} & f_{2,2} \end{bmatrix} \qquad (2.75)$$

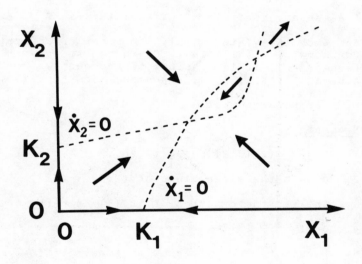

Figure 2.17. Phase-plane plot for model (2.74) when there are intersections of the isoclines along which the $dx_i/dt = 0$.

with the $f_{i,j}$ again as in Section 2.3. For this

2 x 2 Jacobian matrix, as for others, if we define:
(i) the "trace" of the matrix as

$$\text{Tr}(A) = \Sigma_i a_{ii} \qquad (2.76)$$

where the a_{ii}'s are the diagonal elements of the

matrix, and (ii) define the "discriminant" of the
matrix as

$$D(A) = [\text{Tr}(A)]^2 - 4|A| \qquad (2.77)$$

then the eigenvalues of the matrix are given by

$$2\lambda_\pm = \text{Tr}(A) \pm \sqrt{D(A)} \qquad (2.78)$$

Local asymptotic stability therefore requires
$\text{Tr}(A) < 0$ and $|A| > 0$. In the case of (2.75), we
have

$$\text{Tr}(A) = f_{1,1} + f_{2,2}$$

and

$$|A| = f_{1,1}f_{2,2} - f_{1,2}f_{2,1}$$

A necessary condition for stability is $f_{i,i} < 0$

for both i . This is only the start of analysis
for models of this kind. Given additional
assumptions, it is possible to proceed farther (cf.
Albrecht et al., 1974; Freedman, 1980, pp.175-176).

2.6 EXERCISES

Elementary

1. Derive (2.18.2) using (2.20).

2. Obtain the eigenvalues of (2.19).

3. Analyse the dynamics of (2.15) near the origin.

4. Derive (2.24).

5. For system (2.15), show that asymptotic
stability of $\underline{x}_2*$ requires $\beta_2 K_1 > \alpha_2$.

6. Show that (2.30) satisfies (2.29).

7. Derive (2.31).

8. Show that $\underline{x}_2*$ from (2.27) is unstable if $\alpha_2 > K_1 \beta_2$.

9. Obtain the special form of (2.72) at interior
equilibrium.

Intermediate

10. Find the limit of the LHS of (2.13) when $x_2 \to \infty$.

11. Analyse the dynamics of (2.15) near $\underline{x}_3^*$.

12. Show that $\underline{x}_4^*$ from (2.30) is unstable if $\beta_i K_j > \alpha_i$, i not equal to j , for the two possible pairings of i and j .

13. Derive (2.64).

14. Derive (2.68).

Advanced

15. Show that model (2.47) cannot have limit cycle solutions.

Chapter Three

PREDATION

Predation is one of the most obvious ways in which
one population can influence another. Wholly
destructive predation, in which one long-toothed
animal gobbles up another, is only the most obvious
form of ecological interaction in which the
benefits are strictly asymmetrical. From the
theoretical standpoint, such predation is also
equivalent to grazing, parasitism, and infection,
in so far as they depress "victim" population
levels while enhancing those of the "exploiter".
Such generalized predation is perhaps the most
important factor linking together the populations
that compose ecosystems.
    Here we shall be treating the interaction
between just one "prey" species and one "predator"
species. We assume that increasing predator
densities depress the rate of population growth of
the prey species, while increasing prey densities
enhance the rate of popuation growth of the
predator species. Whenever such a relationship
between the population dynamics of two species
exists, we will be treating it as an instance of
predation.

3.1   LOTKA-VOLTERRA MODELS

As before, we begin with the most convenient
mathematical assumptions: continuous-time with
linear functions giving the dependence of the rates
of increase on the interactions between predator
and prey species. These are of course the
assumptions that Lotka and Volterra used, as
discussed before. We will be taking x to
represent the prey population density and y to
represent the predator population density.

Original Lotka-Volterra Model

We begin with the same model that Volterra did, one in which it is assumed that there are no intraspecific density-dependent effects (Scudo and Zeigler, 1978), so that the only density-dependent regulating factor is the density of the other species in the two-species ecosystem. This assumption, together with the preceding ones, immediately gives the following model:

$$dx/dt = x(\alpha - \beta y) \qquad (3.1.1)$$
$$dy/dt = y(\delta x - \sigma) \qquad (3.1.2)$$

We shall take all parameters strictly positive: β and δ because we must in order to represent predation; α because otherwise $x(t)$ goes to 0 as t proceeds to ∞ unless $\alpha = y(0) = 0$; and σ because otherwise $y(t)$ goes to ∞ as t goes to ∞ for $x(0) > 0$ and $y(0) > 0$. Obviously, there is the usual equilibrium at the origin, which is unstable, because of our assumption about α . [Convince yourself of this algebraically if it is not at once obvious to you.]

   Since the zero-isoclines for dx/dt and dy/dt must be linear, due to the Lotka-Volterra assumptions, we know that they will have at most one intersection in the first quadrant of the phase-plane. In fact, one isocline is parallel to the x-axis and above it, that for dx/dt , while the other, for dy/dt , is parallel to the y-axis and to the right of it. This guarantees that there will in fact always be exactly one intersection of the isoclines, and therefore ensures a unique admissible interior equilibrium. From solution of equations (3.1), this equilibrium must be given by

$$(x^*, y^*) = (\sigma/\delta, \alpha/\beta) \qquad (3.2)$$

If we consider the Jacobian matrix about (x^*, y^*) we find

$$\begin{bmatrix} \alpha - \beta y^* & -\beta x^* \\ \delta y^* & -\sigma + \delta x^* \end{bmatrix} = \begin{bmatrix} 0 & -\beta\sigma/\delta \\ \alpha\delta/\beta & 0 \end{bmatrix} \qquad (3.3)$$

This Jacobian matrix has the following eigenvalues:

$$\lambda_{\pm} = \pm\sqrt{-4\sigma\alpha}/2 = \pm i\sqrt{\sigma\alpha} \qquad (3.4)$$

These eigenvalues are purely imaginary complex numbers; their real parts are exactly zero. Recall from Chapters One and Two that the real parts of these eigenvalues give the nature of the flow about the equilibrium defined by the eigenvectors associated with these eigenvalues. When the real parts are all negative, the system flows toward the equilibrium, making it locally asymptotically stable. When one or more of the real parts are positive, then the flow is away from the equilibrium, and the equilibrium is unstable. When neither condition obtains, we have "neutral stability", in which the system flow near the equilibrium does not exhibit any particular tendency to convergence or divergence. Thus there are no asymptotically stable equilibria to which the system state converges.

Fortunately, it is possible to find explicit solution trajectories for the population dynamics with model (3.1). As we did for the first competition and population growth models, we divide (3.1.2) by (3.1.1) and use separation of variables and integration:

$$dy/dx = y(\delta x - \sigma)/[x(\alpha - \beta y)]$$

giving

$$(\alpha - \beta y)dy/y = (\delta x - \sigma)dx/x$$

giving

$$\int_{y(0)}^{y(t)} v^{-1}(\alpha - \beta v)dv = \int_{x(0)}^{x(t)} u^{-1}(\delta u - \sigma)du$$

giving

$$\alpha \ln[y(t)/y(0)] - \beta[y(t)-y(0)]$$

$$= -\sigma \ln[x(t)/x(0)] + \delta[x(t)-x(0)]$$

(3.5)

Numerical solutions of system trajectories using equation (3.5) show that if arbitrary initial conditions, $[x(0),y(0)]$, are chosen, then for some time $t > 0$, say t_1 , $x(t_1) = x(0)$ and $y(t_1) = y(0)$. Similarly, starting from $x(t_1)$ and $y(t_1)$ there must be some t_2 such that $x(t_2) = x(t_1) = x(0)$ and $y(t_2) = y(t_1) = y(0)$, because there is nothing specific to any $y(0)$ which gives rise to the first cycle. It can be proved formally that this pattern of cyclical recurrence is true without exception for all strictly positive $x(0)$ and $y(0)$ (Freedman, 1980, pp.35-36).

The proof hinges on the use of Liapunov functions, which therefore deserve some introduction. [See Hirsch and Smale (1974, pp.192-199) for a rigorous introduction to Liapunov functions.] Here we will be quite informal. Say we have a dynamical system involving a vector

state-variable, $\underline{x}$, and a vector function of $\underline{x}$
giving $d\underline{x}/dt$:
$$d\underline{x}/dt = f(\underline{x})$$
where $d\underline{x}/dt$ is to be understood as the vector of
individual dx_i/dt values. Suppose that there is

an $\underline{x}^*$ such that $f(\underline{x}^*) = 0$, so that $\underline{x}^*$ is an
equilibrium. Liapunov theory is based on the
following ideas. If the equilibrium point
operates like the bottom of a round-bottomed bowl,
with the system-state trajectory acting like the
path of a small ball rolling to the bottom of such
a bowl, then the equilibrium is globally
asymptotically stable. This pattern of declining
deviation from equilibrium will arise when it is
possible to define an analogue of a potential
energy function which steadily declines along
system-state trajectories, hitting zero only at the
equilibrium. Thus, crudely formulated, Liapunov's
Theorem is as follows:
IF: For $\underline{x}$, f , and $\underline{x}^*$ as above, we have a
scalar function V of $\underline{x}$ which is continuous and
smooth with (i) $V(\underline{x}^*) = 0$, (ii) $V(\underline{x}) > 0$ for $\underline{x}$
not equal to $\underline{x}^*$, and (iii) $dV/dt \leq 0$ along
solution trajectories,
THEN: (A) $\underline{x}^*$ is stable; (B) if $dV/dt < 0$ then
$\underline{x}^*$ is asymptotically stable; and (C) if $dV/dt = 0$, for all points on the phase-plane, then all
trajectories are periodic with equations given by
the family of equations $V(\underline{x}) = c_o$, where c_o is a

constant which depends on the initial conditions.
    One aspect of the statement of this theorem
will probably be unfamiliar: the method of
differentiating a function along a trajectory. For
vector systems, differentiating along a trajectory
is performed using the following formula.
$$dV[\underline{x}(t)]/dt = \Sigma_i\{\partial V[\underline{x}]/\partial x_i\big|_{\underline{x}=\underline{x}(t)}\}f_i(\underline{x}) \quad (3.6)$$
where $dx_i/dt = f_i(\underline{x})$, V is a scalar function

of $\underline{x}$, and $\underline{x}$ is a vector. In the two-component
vector case, the one of interest here,
$$dV[\underline{x}]/dt = \{\partial V[\underline{x}]/\partial x_1\}f_1(\underline{x}) + \{\partial V[\underline{x}]/\partial x_2\}f_2(\underline{x})$$

The interpretation of this mathematical operation
is that since V is a scalar function of the
variables x_1 and x_2 it can be thought of as the

dimension of height associated with position on a
landscape, and differentiating V along a

trajectory amounts to assessing whether or not
height is decreasing as position changes. Thus, if
we were studying the flow of water, and V was
height of the water level, then V would steadily
decrease along the course of a mountain stream,
because water level always falls, due to the action
of gravity. In a sense, we are ascertaining
whether or not there is some analogue of this
physical process in our population dynamics.

For system (3.1), there is no such uniform
decrease in V along system trajectories. Rather,
there is a V such that $dV[\underline{x}]/dt = 0$ for all
values of $\underline{x}$, the formula for this function being

$$\delta\mu - \sigma\ln[1 + \delta\mu/\sigma] + \beta\phi - \alpha\ln[1 + \beta\phi/\alpha] = c_o \qquad (3.7.1)$$

where

$$\mu = x - \sigma/\delta \qquad (3.7.2)$$
$$\phi = y - \alpha/\beta \qquad (3.7.3)$$

and

$$c_o = \delta[x(0) - \sigma/\delta] - \sigma\ln[\delta x(0)/\sigma]$$
$$+ \beta[y(0) - \alpha/\beta] - \alpha\ln[\beta y(0)/\alpha] \qquad (3.7.4)$$

[In this case, equation (3.7.1) is obtained by
re-arranging solution trajectory (3.5).] Given the
stipulations of the Liapunov theorem, showing that
(3.7.1) differentiated according to (3.6) is always
zero, as is to be done in the Exercises, in turn
demonstrates that all solution trajectories are
closed orbits. This type of dynamics is shown in
Figure 3.1.

This pattern of system behaviour is extremely
odd, if it is to be taken seriously as that of a
prospective model of actual ecosystems. The
problem is that real ecosystems are always subject
to some degree of perturbation, however strong the
pattern of density-dependence affecting the
constituent species. Normally this isn't a problem
if the asymptotic behaviour of the system is
determined by attractors, be they equilibria, limit
cycles, or strange attractors. But perturbations
to trajectories like those indicated in Figure 3.1
are not damped. Instead, the trajectory simply
follows a new periodic orbit after removal of the
perturbing factor. Thus, in models with such
dynamics, our expected dynamics in the modelled
ecosystem could be as shown in Figure 3.2.

In fact, it is doubtful that model (3.1) has
any relevance to the material world of living
predators and their prey, in spite of its
importance in the early development of theoretical
ecology. The reasons for this can be made
mathematically explicit. Consider an infinitesimal
perturbation to model (3.1), as follows:

102

$$dx/dt = x(\alpha - \beta y - \varepsilon x) \qquad (3.8.1)$$
$$dy/dt = y(\delta x - \sigma) \qquad (3.8.2)$$

where $1 \gg \varepsilon > 0$. In addition to the origin, this system has an equilibrium at

$$(x^*, y^*) = (\sigma/\delta, \alpha/\beta - \varepsilon\sigma/\delta\beta) \qquad (3.9)$$

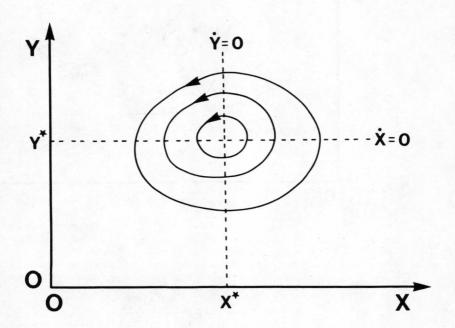

Figure 3.1. Qualitative pattern of
solution trajectories for model
(3.1). The point in the
middle of the trajectories is the neutrally
stable equilibrium point $(\sigma/\delta, \alpha/\beta)$.

Evidently, as ε approaches 0 , equilibrium (3.9)
approaches equilibrium (3.2) arbitrarily closely,
while model (3.8) similarly approaches model (3.1).

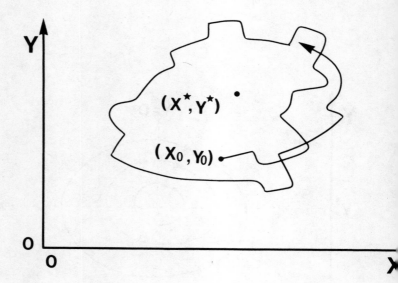

Figure 3.2. Dynamics of a
two-variable model with only
neutrally stable periodic orbits
when subject to intermittent
perturbation.

Now the Jacobian for model (3.8) near equilibrium
(3.9) is

$$\begin{bmatrix} \varepsilon\sigma/\delta\beta - 2\varepsilon & -\beta\sigma/\delta \\ \\ \alpha\delta/\beta - \varepsilon\sigma/\beta & 0 \end{bmatrix} \tag{3.10}$$

which similarly can be made arbitrarily close to

matrix (3.3). Recall from Section 2.5 that two-by-two matrices have eigenvalues given by λ in

$$2\lambda_{\underline{+}} = \text{Tr}\{J\} \underline{+} \sqrt{D}$$

where, in this case,

$$D = [\text{Tr}\{J\}]^2 - 4|J|$$
$$= O(\epsilon^2) - 4[\alpha\sigma + O(\epsilon)]$$
$$= -4\alpha\sigma + O(\epsilon) \leq 0$$

Therefore, the only Real components of the λ's are given by the $\text{Tr}\{J\}$ term, which is

$$\epsilon\sigma/\delta\beta - 2\epsilon = \epsilon(\sigma/\delta\beta - 2) \qquad (3.11)$$

From this, we see that an infinitesimal perturbation can give rise to eigenvalues with non-zero Real parts, which will in turn give rise to either divergence or convergence to the equilibrium in its immediate neighbourhood, behaviour entirely different from that of the original model.

Models which are sensitive to perturbations of this kind are referred to as structurally unstable, and used in scientific theory only if very special "conservation" properties are known to hold, such that the possible model structures are severely constrained (cf. Hirsch and Smale, 1974, p.158, pp.304-318). Since such properties are certainly not known in predator-prey systems, model (3.1) and its neutrally stable periodic trajectories can be dismissed from consideration as relevant models of ecosystem dynamics. [See also May (1974, Ch.3).]

An Alternative Lotka-Volterra Model

Consider a relatively straightforward modification of model (3.1), along the lines of model (2.25):

$$dx/dt = x(\alpha - cx - \beta y) \qquad (3.12.1)$$
$$dy/dt = y(-\sigma + \delta x - dy) \qquad (3.12.2)$$

where all parameters are strictly positive. Evidently, the c and d parameters reflect some form of intraspecific interference effect, or effects. This model was first treated by Samuelson (1967). We will follow the lines of Freedman's (1980, pp.37-41) analysis.

The origin has the same local dynamics for model (3.12) as it had for model (3.1), being unstable in both cases. An additional equilibrium arises at

$$(x^*, y^*) = (K_1, 0) = (\alpha/c, 0) \qquad (3.13)$$

The Jacobian of equilibrium (3.13) is

$$\begin{bmatrix} -\alpha & -\beta K_1 \\ 0 & K_1\delta - \sigma \end{bmatrix}$$

with eigenvalues $-\alpha$ and $K_1\delta - \sigma$, the latter

being equivalent to $\alpha\delta/c - \sigma$. The local asymptotic stability of (3.13) depends on the second eigenvalue. If $\sigma c > \alpha\delta$, then the equilibrium is locally stable. This corresponds to a high death rate of predators, a low prey carrying capacity, and meagre predator success in using the few prey that are available at the prey carrying capacity. If $\sigma c < \alpha\delta$, so that the predator is reasonably successful relative to the prey abundance at the latter's carrying capacity, then this equilibrium is unstable.

If equilibrium (3.13) is unstable, there is an interior equilibrium given by
$$x^* = [\beta\sigma + \alpha d]/[\beta\delta + cd] \qquad (3.14.1)$$
$$y^* = [\alpha\delta - \sigma c]/[\beta\delta + cd] \qquad (3.14.2)$$
Locally, this equilibrium has linearized dynamics given by the following Jacobian matrix

$$\begin{bmatrix} \alpha - 2cx^* - \beta y^* & -\beta x^* \\ \delta y^* & -\sigma + \delta x^* - 2dy^* \end{bmatrix}$$

$$= \begin{bmatrix} -cx^* & -\beta x^* \\ \delta y^* & -dy^* \end{bmatrix} \qquad (3.15)$$

the RHS arising because we require
$$\alpha - cx^* - \beta y^* = -\sigma + \delta x^* - dy^* = 0$$
at this interior equilibrium. Jacobian matrix (3.15) has associated eigenvalues given by

$$2\lambda_+ = -cx^* - dy^* \pm \sqrt{(cx^* + dy^*)^2 - 4(cd + \beta\delta)x^* y^*} \qquad (3.16)$$

Since $cd + \beta\delta > 0$, the magnitude of the discriminant term is less than the trace squared, so that $\mathrm{Tr}\{J\} + \sqrt{D}$ is less than zero. Therefore, the Real parts of both eigenvalues are less than zero, ensuring local asymptotic stability of the interior equilibrium, when it is admissible.

Turning to global analysis, we consider the case arising when equilibrium (3.13) is locally stable. The isoclines for model (3.12) can be found readily, as for system (2.25). When $(K_1,0)$ is stable, they have the pattern shown in Figure 3.3. It is apparent from this figure that

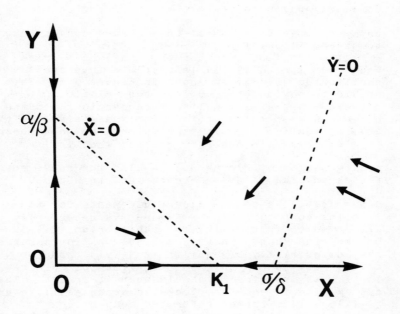

Figure 3.3. Phase-plane portrait of the dynamics of model (3.12) when the isoclines do not intersect in the first quadrant.

$K_1 < \sigma/\delta$ guarantees global convergence to a small neighbourhood of equilibrium (3.13), except for trajectories starting on the y-axis which are not subject to perturbation. For $x > \alpha/c$, we have $dx/dt < 0$. Therefore interior trajectories must eventually reach some time, τ, at which $x(\tau) < \alpha/c$, and similarly for all $t > \tau$. But

$$dy/dt = y(-\sigma + \delta x - dy) \leq y(\delta x - \sigma)$$
so that for $t > \tau$,
$$dy/dt \leq y(\delta\alpha/c - \sigma) \leq y(\delta[\sigma/\delta] - \sigma) = 0$$
from the position of the isocline. This guarantees
convergence to some small neighbourhood of the
x-axis, along which we have
$$dx/dt \approx x(\alpha - cx)$$
a logistic expression, making $x(t)$ converge to
some neighbourhood of $\alpha/c = K_1$, where local

asymptotic stability will then ensure convergence
to the equilibrium.

The two isoclines for the case which arises
when equilibrium (3.13) is unstable intersect at
equilibrium (3.14), as shown in Figure 3.4. This
should be obvious to you from the relationship
between the x-axis-isocline intersections and the
eigenvalues of the Jacobian matrix associated with
equilibrium (3.13). In particular, the requirement
for $K_1 < \sigma/\delta$ in order to get the isocline

configuration of Figure 3.3 is also the requirement
for eigenvalue $K_1\delta - \sigma$ to be less than zero.

Conversely, the parametric requirements for the
isocline configuration of Figure 3.4 necessarily
give rise to instability for equilibrium (3.13).
On the other hand, this same configuration ensures
the admissibility of equilibrium (3.14), and we
have already shown that this equilibrium is always
locally asymptotically stable when it arises. The
system flow is counter-clockwise about equilibrium
(3.14). Unlike the case for the competition
equations studied in the previous chapter, there is
no simple topological argument available to show
global convergence to the unique locally
asymptotically stable equilibrium.

Accordingly, we must analyse the global
dynamics using other methods when the case of
Figure 3.4 arises. One method we can try is to
look for a function meeting the stipulations of
Liapunov's Theorem. Any such function would have
to give global convergence to equilibrium (3.14),
because we cannot have an equilibrium which is
asymptotically stable at least locally in
conjunction with global neutrally stable dynamics,
like that found for the original Lotka-Volterra
predator-prey model. Since model (3.12) is a
generalized version of model (3.1), it is natural
to try a $V(x,y)$ of the same form as (3.6). Let
$$\mu = x - x* \qquad (3.17.1)$$

and
$$\phi = y - y^* \qquad\qquad (3.17.2)$$
where x^* and y^* are as defined by (3.14).

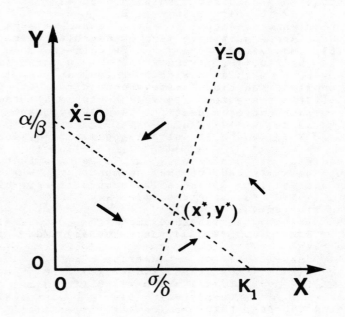

Figure 3.4. Phase-plane portrait of the dynamics of model (3.12) when the isoclines intersect in the first quadrant.

Then we take
$$V(\mu,\phi) = \delta\mu - \delta x^*\ln\{[\mu+x^*]/x^*\}$$
$$+ \beta\phi - \beta y^*\ln\{[\phi+y^*]/y^*\} \quad (3.18)$$
[Note that when $c = d = 0$, equation (3.18) reduces to (3.6).] Evidently, $V(0,0) = 0$, since $\ln\{1\} = 0$. Showing that (3.18) is strictly positive for μ and ϕ not equal to zero is a bit harder. There are two sets of terms of the form
$$Z = z^*\gamma(\Omega/z^* - \ln\{[\Omega+z^*]/z^*\}),$$

where $z*$ is analogous to $x*$ and $y*$, Ω is analogous to μ and ϕ , and Υ is analogous to δ and β . We always have $z*$ and Υ strictly positive, so $Z > 0$ requires

$$\Omega/z* - \ln\{\Omega/z* + 1\} > 0 \qquad (3.19)$$

for Ω not equal to zero . Let

$$u = \Omega/z* + 1$$

so that our required condition becomes

$$W = u - 1 - \ln[u] > 0 \qquad (3.20)$$

for u now not equal to 1 . Since $\Omega > -z*$, for otherwise we must have zero or negative values of $x(t)$ and $y(t)$, we have u within the open interval $(0,\infty)$. At $u = 1$, $W = 0$. We need to show that W is at a global minimum at $u = 1$. Extrema of functions arise when their first derivatives are zero, as you should recall from your mathematics education, thus we evaluate

$$dW/du = 1 - 1/u$$

at $u = 1$, where it uniquely equals zero, as

required. Since $d^2W/du^2 = -u^{-2}$, we know that this must be a minimum, so that we must have $W > 0$ for u not equal to 1 and so we must similarly have $V(\mu,\phi) > 0$ for perturbations away from equilibrium (3.14).

It only remains to examine

$$dV/dt = [\partial V/\partial\mu]d\mu/dt + [\partial V/\partial\phi]d\phi/dt \qquad (3.21)$$

We have

$$\begin{aligned}
d\mu/dt &= dx/dt - dx*/dt \\
&= dx/dt = x(\alpha - cx - \beta y) \\
&= (\mu+x*)[\alpha-c(\mu+x*)-\beta(\phi+y*)] \\
&= (\mu+x*)[-c\mu - \beta\phi] \qquad (3.22.1)
\end{aligned}$$

using the fact that

$$\alpha - cx* - \beta y* = 0$$

and similarly

$$d\phi/dt = (\phi+y*)[\delta\mu - d\phi] \qquad (3.22.2)$$

Using (3.18) and the Chain Rule for differentiation, we also find

$$\begin{aligned}
\partial V/\partial\mu &= \delta - \delta x*(1/x*)x*/[\mu+x*] \\
&= \delta\{1 - x*/[\mu+x*]\} \qquad (3.23.1)
\end{aligned}$$

and similarly

$$\partial V/\partial\phi = \beta\{1 - y*/[\phi+y*]\} \qquad (3.23.2)$$

Assembling all these results together, we obtain

$$\begin{aligned}
dV/dt &= \delta\{(\mu+x*)[-c\mu-\beta\phi] - x*[-c\mu-\beta\phi]\} \\
&\quad + \beta\{(\phi+y*)[\delta\mu-d\phi] - y*[\delta\mu-d\phi]\} \\
&= \delta\mu[-c\mu-\beta\phi] + \beta\phi[\delta\mu-d\phi] \\
&= -c\delta\mu^2 - d\beta\phi^2 < 0 \qquad (3.24)
\end{aligned}$$

for (x,y) not equal to $(x*,y*)$. Thus, from Liapunov's Theorem, we have global asymptotic convergence to $(x*,y*)$ within the interior of the first quadrant. Thus the introduction of any degree of self-limitation of reproductive rates, in the prey or predator population density differential equations, gives rise to systems which converge to global asymptotically stable equilibria, a result radically different from that found for the original Lotka-Volterra model. This further emphasizes the problem of structural instability in that model. In addition, it suggests that predator-prey population dynamics need not be as different from those generated by competition as the impression given by the classical Lotka-Volterra predator-prey model might suggest.

3.2 GENERALIZED PREDATOR-PREY MODELS

One of the crucial assumptions incorporated in model (3.12) is that predation takes an extremely simple form: prey mortality per prey individual is assumed to increase linearly with predator density. Conversely, model (3.12) implicitly assumes that each predator receives a linear increment in net reproductive output as prey density increases, without satiation. The implicit assumptions of this model are easily seen to be absurd. For example, it is assumed that a lion which can eat two zebra/month when there are two hundred zebra should also eat two hundred zebra/month when there are twenty thousand zebra, with a corresponding hundred-fold increase in the lion's reproductive output. However, it seems reasonable to suppose that indigestion might interfere with both prey capture and copulation, at least to some extent. Thus we must allow for saturation effects in the predator's impact on prey population density, and conversely.

Generalized models which can allow for such saturation effects were proposed by Gause (1934) and Gause et al. (1936). Freedman (1980, pp.66-68) has given a somewhat more general form for the Gause model:

$$dx/dt = xg(x) - yp(x) \qquad (3.25.1)$$
$$dy/dt = y[-\gamma + q(x)] \qquad (3.25.2)$$

where:

x and y are as before;
$g(0) > 0$; $dg(x)/dx < 0$;
$g(x)$ is continuous and differentiable;
there exists a $K > 0$ such that $g(K) = 0$;
$p(0) = 0$; $dp(x)/dx > 0$;
$p(x)$ is continuous and differentiable;

$$q(0) = 0 \; ; \quad dq(x)/dx > 0 \; ;$$

and $q(x)$ is continuous and differentiable. Following Freedman, we define:

$$\alpha = g(0) > 0 \qquad\qquad (3.26.1)$$

$$\beta = dp(x)/dx\big|_{x=0} > 0 \qquad (3.26.2)$$

$$\delta = dq(x)/dx\big|_{x=0} > 0 \qquad (3.26.3)$$

$$p_\infty = \lim_{x\to\infty} p(x) \quad \text{in} \quad (0,\infty) \qquad (3.26.4)$$

$$q_\infty = \lim_{x\to\infty} q(x) \quad \text{in} \quad (0,\infty) \qquad (3.26.5)$$

$$K = g^{-1}(0) > 0 \qquad\qquad (3.26.6)$$

where g^{-1} represents the inverse function of g.

Evidently, there is a trivial equilibrium at $(0,0)$. This equilibrium must be locally unstable, because the Jacobian matrix for model (3.25) near the origin has α as one of its eigenvalues, as you are to show in the Exercises. The analysis of the dynamics about $(K,0)$ is also left as an exercise.

The third possible type of equilibrium is an interior equilibrium, (x^*,y^*), with $x^* > 0$ and $y^* > 0$ as well as

$$dx/dt\big|_{(x^*,y^*)} = dy/dt\big|_{(x^*,y^*)} = 0$$

Such an equilibrium need not exist. Firstly, we must have

$$q_\infty > \gamma \qquad\qquad (3.27.1)$$

for otherwise, since $q(x) \leq q_\infty$ because $dq(x)/dx > 0$,

$$dy/dt = y[-\gamma+q(x)] \leq y[-\gamma+q_\infty] < 0$$

for $y > 0$, forestalling the possibility of $y^* > 0$. Henceforth, we shall assume that (3.27.1) holds, and thus we infer the existence of an x^* such that

$$q(x^*) = \gamma \qquad\qquad (3.27.2)$$

The fact that q is continuous and $dq(x)/dx > 0$ guarantees that this x^* is unique. [You might check this for yourself by drawing a variety of $q(x)$ functions conforming to our present assumptions and seeing if you can contrive one in which more than one value of x can equal a given value of q.] Evidently, $dy/dt\big|_{x=x^*} = 0$. The existence of a y^* such that that both differential equations are identically zero now depends only on (3.25.1) at x^*. We require

$$y^* = x^*g(x^*)/p(x^*) \qquad (3.27.3)$$

such that $y^* > 0$. However, it is possible that x^* defined by (3.27.2) is an $x^* > K$, so that $g(x^*) < 0$, giving rise to $x^*g(x^*) < 0$ and thus $y^* < 0$. [Recall that $p(0) = 0$ while $dp(x)/dx > 0$, so that we must have $p(x^*) > 0$.] Thus the condition $y^* > 0$ implicitly requires

$$x^* < K \qquad (3.27.4)$$

Finally, you should note that the uniqueness of x^* and the fact that both g and p functions have only one value for each value of their shared argument ensure that y^* will also be unique. Therefore, when an interior equilibrium meeting conditions (3.27) arises, it will be unique.

Reverting to model (3.25), it turns out that the Jacobian matrix at (x^*, y^*) has the following form

$$\begin{bmatrix} b(x^*) & -p(x^*) \\ y^*dq(x^*)/dx & 0 \end{bmatrix} \qquad (3.28.1)$$

where

$$b(x^*) = x^*dg(x^*)/dx + g(x^*)$$
$$- x^*g(x^*)[p(x^*)]^{-1}dp(x^*)/dx \qquad (3.28.2)$$

as is to be shown in the Exercises. The eigenvalues of this matrix are given by

$$2\lambda_+ = b(x^*) \pm \sqrt{b(x^*)^2 - 4y^*p(x^*)dq(x^*)/dx} \qquad (3.29)$$

From the fact that y^*, $p(x^*)$, and $dq(x^*)/dx$ are all strictly positive, it is clear that the expression under the square-root sign is either

negative or less than $b(x^*)^2$. Thus the signs of the real parts of the eigenvalues depend on the sign of $b(x^*)$ only, giving local asymptotic stability if $b(x^*)$ is less than zero and instability if $b(x^*)$ is greater than zero. [The case of $b(x^*) = 0$ is of mathematical interest only, in that the g, p, or q functions which give rise to this result can be approached arbitrarily closely by functions which do not.]

As was the case for the generalized competition model in continuous-time, the stability condition on $b(x^*)$ can be given a geometrical interpretation. Firstly, note that, from (3.25.1), the prey zero-isocline on which $dx/dt = 0$ is given by

$$y = xg(x)/p(x) \qquad (3.30)$$

Secondly, since x* and g(x*) are both strictly positive, the sign of b(x*) is the same as that of b(x*)/[x*g(x*)] . Thirdly, from (3.28.2),

$$b(x)/[xg(x)] = [dg(x)/dx]/g(x) + 1/x$$
$$- [dp(x)/dx]/p(x)$$
$$= d\{ln[xg(x)/p(x)]\}/dx \qquad (3.31)$$
$$= \{p(x)/[xg(x)]\}d\{xg(x)/p(x)\}/dx$$

since

$$d\{ln[f(x)]\}/dx = \{[f(x)]^{-1}\}df(x)/dx$$

We know that p(x) is strictly positive, so the sign of b(x*) is the same as the sign of d{x*g(x*)/p(x*)} . But this last expression is simply the slope of the dx/dt isocline where it meets the dy/dt isocline. Thus we can infer that the interior equilibrium is unstable when the dx/dt isocline slopes upward at the equilibrium, and conversely the interior equilibrium must be stable when the dx/dt isocline slopes downward at the equilibrium. This surprisingly simple graphical stability criterion was first found by Rosenzweig and MacArthur (1963), although their discussion of its applicability was not valid (cf. Freedman, 1980, p.75).

Global properties of predator-prey models like that of (3.25) have been analysed in considerable depth by Albrecht et al. (1974), Bulmer (1976), and Freedman (1980), following the lines of an early suggestion due to Kolmogorov (1936) concerning the application of the Poincaré-Bendixson Theorem (Hirsch and Smale, 1974, Ch.11). [This theorem concerns the existence of attractors in phase-plane dynamical systems.] It has been found that reasonable predator-prey models may have interior, asymptotically stable, equilibria, limit cycles, or both simultaneously. In particular, the lack of stable equilibria in such predation models implies the existence of at least one limit cycle. The eventual fate of the two populations is then indefinite oscillation, following a well-defined orbit. In addition, for models like (3.25), convergence to the interior equilibrium, when it exists and possesses asymptotic stability, is always oscillatory. This may be contrasted with convergence to stable competitive equilibria, where such convergence is without oscillation. Thus there is a general oscillatory tendency which arises from a simple change in the nature of the modelled ecological interaction, from competition to predation. The three main alternatives for the global dynamics of these models are shown in Figures 3.5, 3.6, and 3.7.

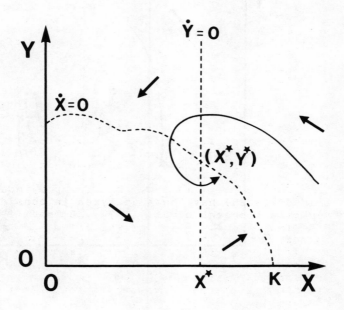

Figure 3.5. Phase-plane for model (3.25) with a global asymptotically stable equilibrium.

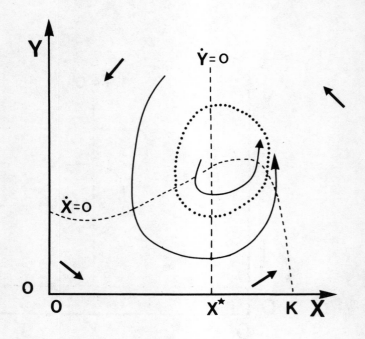

Figure 3.6.   Phase-plane for model
(3.25) with a limit cycle and no
stable equilibria.   There can be more
than one limit cycle, in principle,
nested one within another.

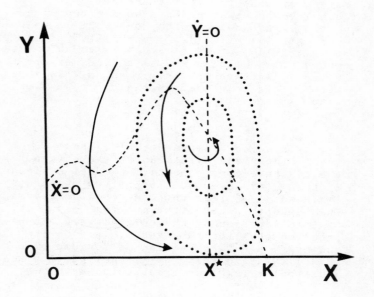

Figure 3.7.   Phase-plane for model
(3.25) with both a limit cycle and an
asymptotically stable equilibrium,
showing the separatrix between their
basins of attraction.   Again, there
may be multiple limit cycles.

3.3   DISCRETE-TIME MODELS

As in the previous two chapters, we proceed from
continuous-time to discrete-time models of the
process of interest. Some of these models will shed
further light on modelling issues which arose in
Section 3.1.

Lotka-Volterra Model without Density-Dependence

Let us return to the problem of the neutral stability of the classical Lotka-Volterra predator-prey model. This model can also be stated in terms of discrete-time equations, with similar linear equations for the population dynamics. We start by assuming no intraspecific density-dependent effects on net reproductive output. By analogy with model (3.1), we write

$$x(t+1) = x(t)[1 + \alpha - \beta y(t)] \qquad (3.32.1)$$
$$y(t+1) = y(t)[1 - \sigma + \delta x(t)] \qquad (3.32.2)$$

where all variables are as before, except that $1 \geq \sigma$. We have the origin as an equilibrium, as before. Again, it is unstable for $\alpha > 0$, because with $x(t) \approx y(t) \approx 0$,

$$x(t+1) \approx x(t)[1 + \alpha]$$

and

$$y(t+1) \approx y(t)[1 - \sigma]$$

giving rise to eigenvalues $1+\alpha$ and $1-\sigma$, for the local linearized system about the origin. As for the discrete-time competition model's linearized systems, for asymptotic stability we require both $1+\alpha$ and $1-\sigma$ less than one in absolute magnitude. This condition cannot be met for $\alpha > 0$.

In addition to the origin, we again have the interior equilibrium given by (3.2), with $x^* = \sigma/\delta$ and $y^* = \alpha/\beta$. Proceeding as for Jacobian matrix (2.58), we find that the linearized flow near this equilibrium is given by Jacobian matrix

$$\begin{bmatrix} 1 & -\sigma\beta/\delta \\ \delta\alpha/\beta & 1 \end{bmatrix} \qquad (3.33)$$

with eigenvalues

$$\lambda_{\pm} = 1 \pm i\sqrt{\sigma\alpha}$$

so that

$$|\lambda_{\pm}| = \sqrt{1 + \sigma\alpha} > 1$$

necessarily giving rise to local instability of equilibrium (3.2) for system (3.32). Thus, unlike the neutral stability of model (3.1), the analogous equilibrium for model (3.32) is strictly unstable. This again suggests the scientific irrelevance of the neutrally stable cycles of model (3.1), in that they do not arise at all in the analogous discrete-time model.

Lotka-Volterra Model with Density-Dependence

By analogy with the continous-time

density-dependent variant of the original
Lotka-Volterra predation model, we write
$$x(t+1) = x(t)[1+\alpha - cx(t) - \beta y(t)] \quad (3.34.1)$$
$$y(t+1) = y(t)[1-\sigma + \delta x(t) - dy(t)] \quad (3.34.2)$$
where at least one of the c or d parameters is
greater than zero, and both are non-negative. The
origin in this model has the same local dynamics as
those of model (3.32). When $c > 0$, equilibrium
point (3.13) again arises, with
$$(x^*,y^*) = (\alpha/c,0) = (K_1,0)$$

The Jacobian matrix for this equilibrium is now

$$\begin{bmatrix} 1 - \alpha & -K_1\beta \\ 0 & 1 - \sigma + \delta K_1 \end{bmatrix}$$

with eigenvalues $1 - \alpha$ and $1 - \sigma + \delta K_1$. We
have $1 \geq \sigma$, so that
$$1 - \sigma + \delta K_1 > 0$$
For
$$1 > |1 - \sigma + \delta K_1|$$

we need $\sigma > \delta K_1$. We already have $\alpha > 0$, so

that for $1 > |1 - \alpha|$, we require $2 > \alpha$. Thus
asymptotic stability of the boundary equilibrium
requires $\sigma > \delta K_1$, as in the continuous-time case,

together with $2 > \alpha$. Having the parameter α
much larger than 1 corresponds to an extremely
high rate of net reproduction, an assumption which
is probably rarely warranted for most prey.
    As before, there is a third possible
equilibrium, as given by equation (3.14), which for
the present model has Jacobian matrix

$$\begin{bmatrix} 1 - cx^* & -\beta x^* \\ \delta y^* & 1 - dy^* \end{bmatrix} \quad (3.35)$$

This Jacobian matrix has eigenvalues given by

$$2\lambda_{\pm} = 2 - (cx^* + dy^*) \pm \sqrt{(2-cx^*-dy^*)^2 - 4\{\beta\delta x^*y^* + (1-cx^*)(1-dy^*)\}} \quad (3.36)$$

Consider the special case with $d = 0$, as may be
reasonable for asocial predators in which there is
no interference competition. Then local asymptotic

119

stability requires
$$2 - cx^* = 2 - c\sigma/\delta < 0 \qquad (3.37.1)$$
and
$$1-cx^* + \beta\delta x^* y^* = 1 +$$
$$[\sigma/\delta][\alpha\delta-c(1+\sigma)] > 0 \qquad (3.37.2)$$
using the conditions on the trace and determinant
of a Jacobian matrix from the end of Chapter 2.
The only parameter in these two inequalities which
plays an unambiguous role in the determination of
stability is α , in that when it is large the
second inequality is more readily met.
Interestingly, a large α destabilizes the
equilibrium at $(K_1,0)$. The other parameters need

to be in an appropriate intermediate range for the
interior equilibrium to be stable. The only
biological interpretation that can be offered for
these results then is that a sufficient capacity
for increase on the part of the prey will foster
the maintenance of predator and prey species in
equilibrium proportions.

Further analysis of models having the form of
(3.34) must rely upon numerical solutions to a fair
degree. However, one principle of model analysis
which can be used here is the consideration of
boundary cases as a guide to the full range of
possible system behaviour. From our treatment of
the discrete-time logistic model, in Section 1.3,
it is evident that discrete-time models with such
second-order terms can have limit cycles and
strange attractors, giving rise to a complex array
of possible system behaviour. This in turn
represents a substantial contrast with the
behaviour of the analogous system (3.12), with its
global asymptotically stable interior equilibrium.
How one interprets this is not very clear. Both
systems are capable of convergence to
asymptotically stable equilibria, so their
respective consequences are not absolutely
distinct, irrespective of parameter values. It
could be argued that the critical parameters of the
discrete-time model will always be such as to allow
essential congruence in the behaviour of the two
types of model, though we will not be considering
such arguments here.

Other Discrete-Time Predation Models

Other discrete-time predation models can be
analysed along the same lines as discrete-time
competition models, using Jacobian matrices for
linear dynamics about equilibria and isocline

intersection patterns to aid in the interpretation
of global dynamics. Unfortunately, it is very
difficult to extract general principles from
broadly formulated discrete-time predation models.

3.4 PARASITOID MODELS

A General Model

One exception to this general rule of opacity in
generalized predator-prey models is provided by the
interactions of univoltine insect parasitoids and
their univoltine hosts. Both of these types of
species have one generation a year, and any given
host larva must die, produce a host adult, or
produce a parasitoid adult. The adults in turn are
often non-feeding. This makes the basic dynamical
properties of the population densities of such
organisms extremely elegant. The general equations
for such host-parasitoid systems are readily
written down, following Hassell (1978, p.9):

$$N_{t+1} = \lambda N_t f[N_t, P_t] \qquad (3.38.1)$$

$$P_{t+1} = c N_t \{1 - f[N_t, P_t]\} \qquad (3.38.2)$$

where N and P give the densities of the host
and parasitoid populations respectively, λ and c
give the reproductive rates of the unparasitized
hosts and the parasitoids produced per infested
host respectively, and $f[\ ]$ gives the probability
that a host will escape parasitism. In general, λ
and c may be functions of N, although commonly
c is taken as unity, since parasitoids can often
survive most sources of host mortality, such as
starvation, disease, further parasitism, and so on.
[Obviously, bird predation would be an exception to
this general rule.]

Taking the subsidiary cases with $c = 1$ and
λ a constant, we analyse the local dynamics about
equilibria, setting $N_{t+1} = N_t = N*$ and
likewise for $P*$:

$$N* = \lambda N* f[N*, P*]$$
$$P* = N* \{1 - f[N*, P*]\}$$

giving

$$f[N*, P*] = 1/\lambda \qquad (3.39.1)$$

and

$$P* = N* \{1 - 1/\lambda\}$$

or

$$\lambda P* = N* [\lambda - 1] \qquad (3.39.2)$$

Admissibility of interior equilibria then requires
$\lambda > 1$. [This is immediately obvious from the fact
that $\lambda < 1$ gives N_t going to zero as time

increases without limit.]

Now we consider the linearized dynamics about equilibria of form (3.39). Rather than using the Jacobian formulation, in this special case we will use the Taylor expansion for small perturbations about equilibrium, following Hassell (1978, pp.201-204). Define the perturbation vector with elements x and y to satisfy the following equations:

$$N_t = N*[1+x_t] \qquad (3.40.1)$$

$$P_t = P*[1+y_t] \qquad (3.40.2)$$

We also take

$$f[N_t,P_t] = f[N*,P*] + x_t N* \partial f[N*,P*]/\partial N$$
$$+ y_t P* \partial f[N*,P*]/\partial P + O(x^2,xy,y^2)$$
$$\approx \{1+\nu x_t - \eta(\lambda-1)y_t\}\lambda^{-1} \qquad (3.41.1)$$

where we take

$$\nu = \lambda N* \partial f[N*,P*]/\partial N \qquad (3.41.2)$$

and

$$\eta = -N* \partial f[N*,P*]/\partial P \qquad (3.41.3)$$

Using (3.38) to (3.41) repeatedly, it is to be shown in the Exercises that the linearized dynamics about (3.39) are well-approximated by

$$\begin{bmatrix} x_{t+1} \\ y_{t+1} \end{bmatrix} = \begin{bmatrix} 1+\nu & -\eta(\lambda-1) \\ \{\lambda-(1+\nu)\}/(\lambda-1) & \eta \end{bmatrix} \begin{bmatrix} x_t \\ y_t \end{bmatrix} \qquad (3.42)$$

As before, we analyse the dynamical properties of system (3.42) by seeking its eigenvalues. These are given by the θ roots of

$$\theta^2 - (\eta+\nu+1)\theta + \eta\lambda = 0 \qquad (3.43.1)$$

which in turn are given by

$$2\theta_+ = \eta+\nu+1 \pm \sqrt{(\eta+\nu+1)^2 - 4\lambda\eta} \qquad (3.43.2)$$

As for discrete-time systems generally, local asymptotic stability of linearized system (3.39) requires $|\theta_+| < 1$. For reasonable parasitoid interactions, we may assume $\lambda > 1$, $\eta \geq 0$, and $\nu \geq 0$. Taking

$$D = (\eta+\nu+1)^2 - 4\lambda\eta$$

where D is simply the discriminant function which we have discussed before, there are two generic cases:

(i) $\underline{D \leq 0}$. In this case, $|\theta_+| = |\theta_-|$, with

$$|\theta_+| = \sqrt{4^{-1}(\eta+\nu+1)^2 + [2^{-1}\sqrt{4\lambda\eta - (\eta+\nu+1)^2}]^2}$$

$$= \sqrt{\lambda\eta}$$

with local asymptotic stability requiring
$$\eta < 1/\lambda$$

(ii) $\underline{D > 0}$. Now the θ_+ are always real with θ_+ not equal to θ_- and

$$2|\theta_+| = |\eta + \nu + 1 \pm \sqrt{(\eta+\nu+1)^2 - 4\lambda\eta}|$$

In this case, $|\theta_+| < 1$ if and only if ("$\underline{iff}$")

$$\sqrt{(\eta+\nu+1)^2 - 4\lambda\eta} < 2 - (\eta+\nu+1)$$

$\underline{iff}$

$$(\eta+\nu+1)^2 - 4\lambda\eta < 4 - 4(\eta+\nu+1) + (\eta+\nu+1)^2$$

$\underline{iff}$

$$\eta + \nu + 1 - \lambda\eta < 1$$

$\underline{iff}$

$$\eta > \nu/(\lambda-1)$$

Thus, taking all cases into account, a sufficient condition for local asymptotic stability for the interior equilibria associated with model (3.38) is

$$1/\lambda > \eta > \nu/(\lambda-1) \qquad (3.44)$$

Hassell (1978, p.204) discusses the global convergence properties of equilibrium (3.39) when condition (3.44) is met. He suggests that, for reasonable $f[N,P]$ functions, condition (3.44) also guarantees global convergence to equilibrium (3.39) from interior initial conditions, and all numerical examples that he has examined have borne this conjecture out.

Classical Nicholson-Bailey Model

The simplest parasitoid model which has been seriously proposed is that of Nicholson and Bailey (1935). In a sense, it is a special case of the classical Lotka-Volterra model for predator-prey population dynamics (cf. Hassell, 1978, Appendix II). In particular, it is assumed that neither prey nor predator has self-limiting density-dependence and, as for model (3.1), the number of parasitoid attacks on hosts is in proportion to their population densities:

$$N_a = aN_t P_t \qquad (3.45)$$

where N_a gives the number of prey attacked and "a" gives the probability that a given parasitoid will attack a given host.

However, though (3.45) is similar to the xy terms of model (3.1), it leads to distinctly different consequences for the population dynamics of the two species, depending on how parasitoid attacks are distributed over the hosts. Most importantly, parasitoids may attack a single host more than once. Only one attack is sufficient to kill the host eventually, while at most one parasitoid is assumed to emerge from each parasitized host. [This is not true in general of parasitoid species, but it is true of many of them.] Thus the relevant parameters for the population dynamics are not given by the frequency distribution of levels of parasitism, but by the number of hosts left unparasitized and the number parasitized. Depending on assumptions about parasitoid attack behaviour, different formulae for these numbers may be obtained. Much of Hassell (1978) is devoted to a discussion of such alternative formulae, treated above in terms of a general f[N,P] giving the number of host individuals which escape parasitism.

Nicholson and Bailey assumed that parasitoid attacks were strictly independent of one another, giving rise to a Poisson process for parasitism. [See Feller (1968, pp.156-164) for an introduction to Poisson processes.] With the average number of parasitoid attacks per host given by N_a/N_t , or aP_t from equation (3.45), it is possible to obtain all the elements of the Poisson distribution. In particular, the probability of complete escape from attack is given by $\exp(-aP_t)$, so that the number of parasitized hosts is given by

$$N_p = N_t[1-\exp(-aP_t)] \qquad (3.46)$$

Using this result to specify f[ ] in model (3.38) and taking c = 1 , we have

$$N_{t+1} = \lambda N_t \exp(-aP_t) \qquad (3.47.1)$$

$$P_{t+1} = N_t\{1-\exp(-aP_t)\} \qquad (3.47.2)$$

Now we endeavour to analyse this model. At equilibrium (3.39), we must have

$$f[N^*,P^*] = \exp(-aP^*) = 1/\lambda$$

and so

$$P^* = a^{-1}\ln(\lambda) \qquad (3.48.1)$$

and

$$N^* = \lambda\ln(\lambda)/\{(\lambda-1)a\} \qquad (3.48.2)$$

Let us consider the local stability properties of this equilibrium. In our previous notation,

$$\nu = \lambda N^* \partial f[P^*]/\partial N = 0 \qquad (3.49.1)$$

while

$$\eta = -N^* \partial f[P^*]/\partial P = -N^*\{-a\}\exp(-aP^*)$$
$$= aN^*/\lambda = \ln(\lambda)/\{\lambda-1\} \qquad (3.49.2)$$

In terms of condition (3.44), we have

$$\eta > \nu/(\lambda-1) = 0$$

because $\eta > 0$ for $\lambda > 1$, and we have taken $\lambda > 1$ already. The other half of the condition is

$$1/\lambda > \ln(\lambda)/(\lambda-1)$$

which is true if and only if

$$1 > \ln(\lambda) + 1/\lambda = g(\lambda)$$

but $g(1) = 1$ and

$$dg/d\lambda = \lambda^{-1} - \lambda^{-2} = \lambda^{-1}(1-\lambda^{-1}) > 0$$

for $\lambda > 1$, so that the second half of the condition cannot be met.

One further problem remains to be resolved, which is showing that we have $D < 0$ in this case. We assume the reverse, taking $D > 0$. Therefore, we must have

$$(\eta+\nu+1)^2 > 4\eta\lambda$$

which is only true if

$$(\eta+1)^2 > 4$$

since we have $\eta > 1/\lambda$, but this in turn is true if and only if

$$\eta = \ln(\lambda)/(\lambda-1) > 1$$

iff

$$h(\lambda) = \ln(\lambda) - \lambda + 1 > 0$$

Note that $h(1) = 0$ while

$$dh/d\lambda = 1/\lambda - 1 < 0$$

for $\lambda > 1$, giving $dh/d\lambda < 0$ and thus $h(\lambda) < 0$, for $\lambda > 1$. Therefore, we must have $D < 0$, and this together with our preceding analysis shows that the interior equilibrium is unstable.

Numerical solutions of system (3.47) suggest that the local instability of equilibrium (3.48) is associated with a lack of other attractor trajectories, such as limit cycles, in the interior of the (N,P) phase-plane (Hassell, 1978, p.6), as illustrated with a numerical example in Figure 3.8. This suggests that the Nicholson-Bailey model is an inadequate model for host-parasitoid interactions, since it seems to predict the eventual extinction of parasitoid or both parasitoid and host, yet there are many stable complexes of parasitoid and host species with univoltine life cycles.

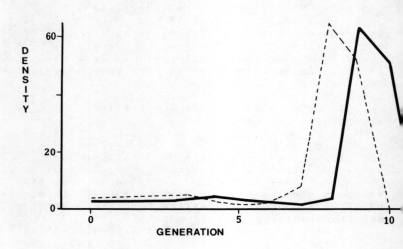

Figure 3.8. Example of a numerical solution trajectory for the Nicholson-Bailey model, $\lambda = 20$, a = 1 , $P* = 3$, $N* = 3.15$, and $N_0 = P_0 = 3$.

Nicholson-Bailey Model with Density-Dependence

The problem of interest, therefore, is to find some alternative model(s) of host-parasitoid population dynamics which at least allow the existence of asymptotically stable equilibria. One such alternative model may be obtained from the Nicholson-Bailey model by making λ density-dependent, as follows:

$$\lambda = \exp\{r[1 - N_t/K]\} \qquad (3.50)$$

Substituting expression (3.50) into model (3.47) gives

$$N_{t+1} = N_t\exp\{r[1 - N_t/K] - aP_t\} \qquad (3.51.1)$$

$$P_{t+1} = N_t[1 - \exp\{-aP_t\}] \qquad (3.51.2)$$

which is a special case of model (8) from Beddington et al. (1976), also discussed in Beddington et al. (1975). Define, for interior equilibrium (N^*,P^*) ,

$$q = N^*/K \qquad (3.52.1)$$

and

$$\phi = r(1-q)/\{1-\exp[-r(1-q)]\} \qquad (3.52.2)$$

From (3.51), it is clear that we must have interior equilibria such that

$$P^* = r(1-q)/a \qquad (3.53.1)$$

and

$$\begin{aligned}N^* &= P^*/[1-\exp\{-aP^*\}]\\ &= r(1-q)/\{a[1-\exp\{-r(1-q)\}]\}\\ &= \phi/a \qquad (3.53.2)\end{aligned}$$

However, the expressions defined by (3.52) make (3.53.2) a complex expression, with the N^* solutions arising as the roots of

$$g(N^*) = r(1 -N^*/K) - aN^*[1-\exp\{-r(1 -N^*/K)\}]$$

If $P^* = 0$, we have the typical boundary equilibria at $N^* = 0$ and $N^* = K$, directly from (3.51.1). Evidently, $(N^*,P^*) = (0,0)$ is unstable if $r > 0$. The stability of the other boundary equilibrium requires more detailed consideration. In the Exercises, it is to be shown that the eigenvalues for the Jacobian matrix about $(K,0)$ are aK and $1-r$, so that local asymptotic stability requires r in the interval $(0,2)$ and a in $[0,1/K)$. The former condition is analogous to the stability constraint arising in the logistic discrete-time model from Section 1.3, as would be expected intuitively.

Finally, we turn to the interior equilibria, as given by (3.53). As is to be shown in the Exercises, the Jacobian matrix about such equilibria is

$$\begin{bmatrix} 1 - rq & -\phi \\ r(1-q)/\phi & \phi - r(1-q) \end{bmatrix} \qquad (3.54)$$

The eigenvalues of this matrix are given by the λ roots of

$$\lambda^2 - \lambda(1-r+\phi) + (1-rq)\phi + r^2q(1-q) = 0 \quad (3.55)$$

As may be readily checked using the roots of this equation, the local dynamics specified by (3.54) often give rise to $(N*,P*)$ points which are locally asymptotically stable in the interior of the phase-plane (N,P). One such instance is that with $r = 1$, $K = 10$, and $a = .1379$, giving $N* = 8$, $P* = 1.45$, as well as eigenvalues

$$\lambda_+ = .552 \pm .276 \, i$$

and

$$|\lambda_+| = |\lambda_-| = .617$$

Convergence to this equilibrium is illustrated in Figure 3.9. Thus, adding density-dependence to the Nicholson-Bailey model leads to local asymptotic stability, unlike the density-independent case. Next we shall consider a case with density-independent population growth, and yet asymptotically stable equilibria.

Generalized Nicholson-Bailey Model

One of the implicit assumptions of the Nicholson-Bailey model is that parasitoids search randomly over the entire habitat. To the extent that parasitoid attack is "clumped", which is to say concentrated in local patches, the Nicholson-Bailey equations do not apply. By way of remedying this, May (1978) has developed an alternative equation for parasitoid reproduction:

$$P_{t+1} = N_t[1 - (1+aP_t/k)^{-k}] \quad (3.56)$$

where a is again as in equation (3.45) and as k approaches ∞ we recover equation (3.46). [Recall the limiting approach to an exponential function from Section 1.3.] The interpretation of k is that, if there are a number of patches, within which parasitism is a Poisson process, but over which different numbers of parasitoids are

distributed with mean π and variance σ_p^2, then

it has been shown (May, 1978, Appendix I) that, to a reasonable approximation,

$$k \approx \pi^2/\sigma_p^2 \quad (3.57$$

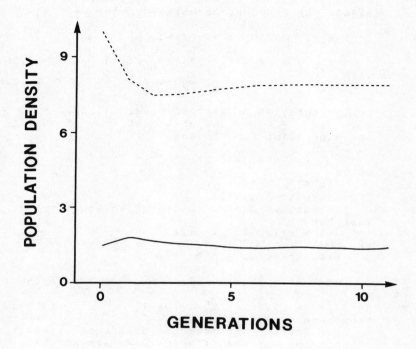

Figure 3.9. Convergence to locally asymptotically stable equilibrium (N*,P*) = (8,1.45) for the parameter values of model (3.51) described in the text.

As k approaches ∞ , there is no variability over patches, and thus they have ceased to exist, giving rise to the original Nicholson-Bailey model. As k approaches 0 , patches are exploited in a highly haphazard fashion, with most patches free of parasitoids. Thus, k can be understood in terms of clumping pattern.

Whatever interpretation is placed on k , and irrespective of the arguments which have been made against its use in this context (e.g., Taylor et al., 1979), equation (3.56) provides an interesting example of the general sort of parasitoid model discussed at the outset of this section. The full system for the host-parasitoid model becomes

$$N_{t+1} = \lambda N_t (1 + aP_t/k)^{-k} \qquad (3.58.1)$$

$$P_{t+1} = N_t[1 - (1 + aP_t/k)^{-k}] \qquad (3.58.2)$$

with

$$f(P_t) = (1 + aP_t/k)^{-k} \qquad (3.58.3)$$

in the notation of general model (3.38). For (3.58), $\nu = 0$ because of the absence of N_t in the expression for f and

$$\eta = -N*\partial f(P*)/\partial P$$

while

$$f(P*) = 1/\lambda$$

from result (3.39.1).

First, we derive the interior equilibrium. We must have

$$(1 + aP*/k)^{-k} = 1/\lambda$$

and thus

$$1 + aP*/k = \lambda^{1/k}$$

so

$$P* = a^{-1}k(\lambda^{1/k}-1) \qquad (3.59.1)$$

and

$$N* = \lambda a^{-1}k(\lambda^{1/k}-1)/[\lambda-1] \qquad (3.59.2)$$

from result (3.39.2).

Second, we consider the asymptotic stability of this interior equilibrium. We already have ν explicitly, while

$$\eta = -N*\{-k(a/k)(1 + aP_t/k)^{-k-1}\}$$

$$= -N*\{-a(1 + aP_t/k)^{-1}\lambda^{-1}\}$$

$$= k(\lambda^{1/k}-1)/\{(\lambda-1)(1 + aP_t/k)\}$$

$$= k(1 - \lambda^{1/k})/(\lambda-1) \qquad (3.60)$$

Turning to local asymptotic stability condition (3.44), it is evident that

$$\eta > 0 = \nu/(\lambda-1)$$

since $\lambda > 1$ and $k > 0$. Thus the only remaining condition on local asymptotic stability is $1/\lambda > \eta$ or

$$1 > \mu(k,\lambda) = k\lambda(1 - \lambda^{-1/k})/(\lambda-1) \qquad (3.61)$$

As k increases, so does the $1 - \lambda^{-1/k}$ term of $\mu(k,\lambda)$, and of course k increases too. Thus $\mu(k,\lambda)$ monotonically increases with k. Now we note that, at $k = 1$,

$$\mu(k,\lambda) = \lambda(1 - 1/\lambda)/(\lambda-1) = 1$$

Thus condition (3.61) is met if and only if

$$k < 1 \qquad\qquad (3.62)$$

This is the sole local asymptotic stability criterion for equilibrium (3.59). Its interpretation is that the parasitoids must attack the hosts in a sufficiently clumped pattern, in effect giving rise to some hosts which largely escape parasitism and considerable parasitoid-parasitoid interference within patches of high parasitoid concentration.

May (1978, p.839) informs us that extensive numerical solutions indicate that global convergence also occurs when condition (3.62) is met, making it a good guide to the properties of system (3.58) as a whole. Hassell (1978, pp.77-78) has generalized model (3.58) to include host density-dependence in either λ or k .

3.5 EXERCISES

Elementary

1. Analyse the stability properties of the origin for model (3.25).

2. Analyse the stability properties of $(K,0)$ for model (3.25).

3. If $x^* > K$, what are the global dynamics of model (3.25)?

4. What are the dynamics of model (3.25) if the y-isocline intersects an x-isocline with a single hump to the left of the maximum of the hump?

5. Derive Jacobian matrix (3.33) and then obtain its eigenvalues.

6. Analyse and explore numerically by iteration some examples of model (3.34). Choose interesting, as opposed to trivial, parameters.

7. Evaluate the behaviour of model (3.47) numerically by iteration of the model equations.

8. Analyse the local dynamics about $(N^*,P^*) = (K,0)$ in the case of model (3.51).

Intermediate

9. Derive Liapunov function (3.7).

10. Show that function (3.7) satisfies the conditions of Liapunov's Theorem, particularly part C.

11. Derive (3.28).

12. Show that there are no asymptotically stable interior equilibria in the predator-prey system given by

$$x(t+1) = x(t)F_1[y(t)]$$

$$y(t+1) = y(t)F_2[x(t)]$$

both F_i being differentiable functions.

13. Derive (3.42).

14. Derive Jacobian matrix (3.54).

Advanced

15. Develop and analyse a host-parasitoid model which incorporates both parasitoid attack "clumping" and density-dependent effects upon host population growth, irrespective of parasitoid mortality.

Chapter Four

SIMPLE ECOSYSTEMS

Up to this point, the models that we have discussed
have been largely inadequate for the purpose of
dissecting out the foundations of ecosystem
organization. While suitably constructed
single-species population growth models can be
excellent predictors of population dynamics, as
well as mathematically convenient, it is a matter
of debate how adequate such models are as devices
for fostering our theoretical understanding of
ecological phenomena.

Two-species models are a great advance in this
respect and, to the extent that they are applied to
species for which a single dyadic interaction is of
paramount importance, they allow fundamental
insights into ecological processes. However, there
are obviously a great many more species which have
a number of important ecological interactions bound
up in the determination of their population growth
rates. Thus we need n-species ecosystem models, of
considerable complexity and generality.

It is easy to write down equations for such
models. It is quite another thing to explore their
range of consequences for ecosystem dynamics.
Unfortunately, the mathematical tools available for
analysing nonlinear dynamical systems of more than
two state variables, such as two species population
densities, are extremely limited. Thus the
available analytical results from ecosystem models
decrease with species number.

We are left with the necessity of
unjustifiably extrapolating from: (a) analysis of
low-species-number models; (b) analysis of special
high-species-number models; and (c) numerical
solutions of complex high-species-number models.
The present discussion is limited to analysis of
types (a) and (b), since approach (c) requires
detailed examination of particular ecosystems.

While there is some validity to the view that the practical dividing line between (a) and (b) occurs just above a species number of 2 , with models having species number greater than 2 being lumped with arbitrary n species models, there are still a few insights to be gained at n = 3 .

This chapter considers such "simple" (i.e., n = 3) ecosystem models. The term "ecosystem" is used here because these models allow multiple interactions. The models of Chapters One to Three were confined to one interaction at most: organism-environment, competitor-competitor, and predator-prey. Here we consider the three possible ecosystem types when n = 3 , under the assumption of no mutualism and more than one trophic level: two predator species feeding on one prey species; one predator species feeding on two prey species; and a three-level food chain. There are of course other types of three-species ecosystem model. Freedman et al. (1983) mention a variety of three-species models involving mutualism. Three competitor models have been discussed by Gilpin (1975), May and Leonard (1975), and Coste et al. (1978). In any case, the present discussion should serve as an introduction to ecological models having this level of complexity.

4.1 TWO PREDATORS AND ONE PREY

Continuous-Time Models
It is possible to discuss continuous-time models of this simple ecosystem in some generality for a peculiar and vitiating reason, as will become clear. Consider the following model:

$$dy_1/dt = y_1 f_1(x) \qquad (4.1.1)$$

$$dy_2/dt = y_2 f_2(x) \qquad (4.1.2)$$

$$dx/dt = xg(x, y_1, y_2) \qquad (4.1.3)$$

where the y_i give the predator species population densities, x gives the prey species density, the f_i are strictly increasing functions of x , and g is a decreasing function of both y_i .

At system (4.1) equilibria which are away from the boundaries of the positive (or first) octant, we require

$$f_1 = f_2 = g = 0$$

To make this more concrete, we plot the f_i functions against x in Figure 4.1. Equilibrium in the interior of the first octant requires that the values of x at which these functions equal zero correspond. Thus, if we write x_i for that value of x at which $f_i = 0$, then we require $x_1 = x_2 = x^*$, together with admissible values of y_1 and y_2 such that $g(x^*, y_1, y_2) = 0$.

But there is no reason, mathematically or ecologically, to expect this to be the case for any particular ecosystem of this type, although it could occur in some specific cases.

Perhaps this point can be made clearer by considering a special case of model (4.1), discussed by Freedman (1980, p. 160):

$$dy_1/dt = y_1[-m_1 + c_1 p_1(x)] \qquad (4.2.1)$$

$$dy_2/dt = y_2[-m_2 + c_2 p_2(x)] \qquad (4.2.2)$$

$$dx/dt = xg(x) - y_1 p_1(x) - y_2 p_2(x) \qquad (4.2.3)$$

where the g and p_i functions are as in model (3.25), and the c_i and m_i parameters are all strictly positive. Model (4.2) has interior equilibria if and only if there exist common solutions, say x^* , to the three equations

$$c_1 p_1(x^*) = m_1 \qquad (4.3.1)$$

$$c_2 p_2(x^*) = m_2 \qquad (4.3.2)$$

$$x^* g(x^*) = y_1 p_1(x^*) + y_2^* p_2(x^*) \qquad (4.3.3)$$

Indeed, there is no reason to expect that such a common x^* will be found even for equations (4.3.1) and (4.3.2), by themselves, to say nothing of (4.3.3), unless there is some unusual relationship between the p_i functions, such as identity or parallel form. [A biological rationale for such similarity might be that the species are congeneric.] Therefore, there are unlikely to be three-species equilibria when two, or more, predators utilize a single common prey species.

Ostensibly, this suggests that predators cannot "share" prey, ecologically.

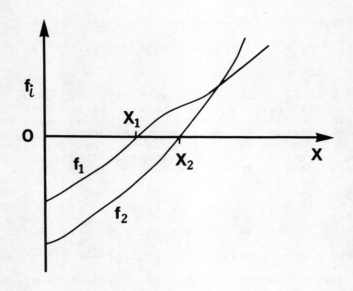

Figure 4.1. Plots of typical predator net reproductive rates as functions of prey density for model (4.1). The x_i indicate the values of x at which the $f_i = 0$.

As has been discussed in the literature ·(Haigh and Maynard Smith, 1972; Armstrong and McGehee, 1980), it is this general problem of equilibrium non-existence which underlies the basic idea of the number of limiting resources limiting the maximum number of resource-exploiter species which can co-exist. In effect, this is another version of the "competitive exclusion" idea. In fact, this interpretation of the theory is too shallow. Haigh and Maynard Smith (1972) show how a variety of realistic ecological complexities could allow all three species to co-exist at an asymptotically stable equilibrium. For our present purposes, the points raised by Armstrong and McGehee (1980) are even more interesting. Proceeding partly from a sophisticated mathematical analysis of model (4.1) (McGehee and Armstrong, 1977), they show that non-existence of an equilibrium does not preclude the global persistence of all three species, starting from the interior of the first octant of (x, y_1, y_2) space.

Their main point is effectively made by considering a single example. Let us discuss the following model:

$$dy_1/dt = y_1(-.1 + .15x/[x+50])$$

$$dy_2/dt = y_2(-.11 + .00099x)$$

$$dx/dt = x\{.1(1 - x/300) - .5y_1/[x+50] - .003y_2\}$$

As stated in Armstrong and McGehee (1980, p.155), this system seems to have at least one interior attractor trajectory, a limit cycle, which involves the persistence of all three species, even though it has no interior equilibria. Thus equilibrium existence, to say nothing of equilibrium stability, is not a requirement for the persistence of such three-species ecosystems. Unfortunately, in the absence of equilibria, ecological models with more than two species are extremely difficult to analyse. Most of our mathematical theory revolves around treatments of system dynamics in the immediate vicinity of equilibria, stable or unstable, even in the case of limit cycles. Those additional mathematical tools which can be deployed for the analysis of the dynamics of "planar" (two-dimensional) models, such as the Poincaré-Bendixson Theorem, do not carry on into higher-dimension models. Thus there is little which can be said about the dynamics of two-predator-one-prey ecosystem models.

More generally, the present case illustrates the problematic nature of theoretical arguments in ecology which identify asymptotically stable equilibria with the persistence of species within an ecosystem (cf. May, 1974). Persistence simply does not require such equilibria, in principle.

Moreover, it is equally evident that at least some analysis of the global dynamics of a model is necessary to unravel its full scientific import; consideration of equilibrium dynamics alone may be misleading. The most important corollary of this is that we generally do not have a sufficient grasp of the possible dynamical features of many of the key models in ecology, as will be discussed in more detail in the next chapter.

Discrete-Time Models: Two Parasitoids

As with the continuous-time theory, the same problem of equilibrium non-existence can arise in discrete-time models in which the two predators have no direct effect upon each other's population dynamics, other than that through indirect competition for prey. Therefore there is little reason to pursue such models here, in general.

On the other hand, there is a biologically well-founded sub-set of such models in which this problem does not arise. Parasitoid species often share hosts, but in so doing undergo intense competition. This competitive interaction can take on a number of extreme forms: (i) two species may oviposit at the same life cycle stage, but one of the two always destroys the other; (ii) one parasitoid species may kill all hosts that it infests before they have the chance to reach the life cycle stage used by a second parasitoid species; or (iii) a parasitoid species which enters the host later in the life cycle consumes those parasitoid larvae which were oviposited earlier, in addition to remaining host tissue. All these ecological interactions are readily embodied in the following model:

$$N_{t+1} = \lambda N_t f_1(P_t) f_2(Q_t) \qquad (4.4.1)$$

$$P_{t+1} = N_t[1 - f_1(P_t)] \qquad (4.4.2)$$

$$Q_{t+1} = N_t f_1(P_t)[1 - f_2(Q_t)] \qquad (4.4.3)$$

with host density N_t, parasitoid species densities P_t and Q_t, as well as probabilities of host parasitism of f_1 and f_2 for parasitoids "P" and "Q", respectively. This model has been discussed by Hassell (1978), Hassell (1979), and, in greatest detail, May and Hassell (1981).

Model (4.4) has been analysed with

$$f_1(P_t) = \exp(-a_1 P_t) \qquad (4.5.1)$$

$$f_2(Q_t) = \exp(-a_2 Q_t) \qquad (4.5.2)$$

thereby assuming completely dispersed parasitoid attack (Nicholson, 1933; Nicholson and Bailey, 1935). The problem with functions (4.5) is that they give rise to a model which generally lacks admissible equilibria with all three species present, and when these equilibria do arise they are unstable (Hassell, 1978, pp.149-150).

Instead, consider these analogues of function (3.58.3):

$$f_1(P_t) = [1 + a_1 P_t / k_1]^{-k_1} \qquad (4.6.1)$$

$$f_2(Q_t) = [1 + a_2 Q_t / k_2]^{-k_2} \qquad (4.6.2)$$

used in May and Hassell (1981). As before, the a_i give the searching efficiencies, while the k_i give the degree of clumping of parasitoid attack. We shall also assume that λ is a density-independent constant greater than one.

Following May and Hassell (1981, Appendix C), we seek equilibrium population densities for the three populations, writing them as N^*, P^*, and Q^*. From equation (4.4.1), it is clear that at equilibrium we must have

$$1 = \lambda f_1(P^*) f_2(Q^*) \qquad (4.7.1)$$

Equation (4.4.2) similarly gives

$$N^* = P^*[1 - f_1(P^*)]^{-1} \qquad (4.7.2)$$

and thus equation (4.4.3) together with these conditions gives

$$Q^* = P^*[1-f_1(P^*)]^{-1} f_1(P^*)[1-\lambda^{-1} f_1(P^*)^{-1}]$$

$$= P^*[f_1(P^*)-\lambda^{-1}][1-f_1(P^*)]^{-1} \qquad (4.7.3)$$

Finally, we find the following expression with only one unknown variable, from which the equilibrium can be found using (4.7.2) and (4.7.3):

$$1 = \lambda f_1(P^*) f_2\{P^*[f_1(P^*)-\lambda^{-1}][1-f_1(P^*)]^{-1}\} \qquad (4.7.4)$$

It is possible to find fairly simple conditions for the occurrence of three-species interior equilibria. Let

$$f_1 = f_1(P*)$$

and notice that

$$a_1 P* = k_1(f_1^{-1/k_1} - 1)$$

Using these two equations, expression (4.7.4) can be re-written as

$$a_2/a_1 = \lambda k_2 [(\lambda f_1)^{1/k_2} - 1](1-f_1)$$
$$+ k_1(\lambda f_1 - 1)(f_1^{-1/k_1} - 1) \qquad (4.8)$$

as is to be shown in the Exercises. From (4.8) and specific choices of the parameters, f_1 can be found numerically, and thus $P*$, $Q*$, and $N*$ in turn. From (4.6.1), it is clear that

$$f_1(P) < 1$$

for $P > 0$, and as P approaches ∞ , $f_1(P)$ approaches 0 . As $P*$ approaches 0 , f_1 approaches 1 , and the RHS of (4.8) becomes

$$\lim_{f \to 1} \{\lambda k_2 [(\lambda f_1)^{1/k_2} - 1](1-f_1)$$
$$+ k_1(\lambda f_1 - 1)(f_1^{-1/k_1} - 1)\}$$
$$= \lambda k_2(\lambda^{1/k_2} - 1)(\lambda - 1)^{-1}\{-1/[k_1(-1/k_1)]\}$$

first factoring out terms which do not involve f_1 and then using L'Hôpital's Rule to obtain the limit,

$$= \lambda k_2(\lambda^{1/k_2} - 1)(\lambda - 1)^{-1}$$
$$= \beta_2(\lambda, k_2) \qquad (4.9.1)$$

in the notation of May and Hassell (1981). As f_2 approaches 1 , and thus $Q*$ approaches 0 , the equation

$$1 = \lambda f_1 f_2$$

guarantees that λf_1 will approach 1 . Thus an analogous calculation of the limiting value of the RHS of equation (4.8) when $Q*$ approaches 0 , which is to be done in the Exercises, gives

$$\beta_1(\lambda, k_1) = (\lambda-1)[k_1(\lambda^{1/k_1} -1)]^{-1} \qquad (4.9.2)$$

It is necessary for interior equilibria that we have both P^* and Q^* strictly positive, so that we must have the LHS of (4.8) such that

$$\beta_2 > a_2/a_1 > \beta_1 \qquad (4.10)$$

An interesting point is whether or not condition (4.10) is sufficient for equilibrium existence, given that it is evidently a necessary condition. When condition (4.10) is satisfied, we have $P^* > 0$ and $Q^* > 0$. All that remains to consider is whether or not $N^* > 0$, which can be addressed using (4.7.2). From (4.7.1), we must have f_1 in the interval $(1/\lambda, 1)$, so that

$$N^* = P^*(1 - f_1)^{-1}$$

is in the interval $(\lambda P^*[\lambda-1]^{-1}, \infty)$. Since $\lambda > 1$ we therefore have

$$N^* > P^* > 0$$

so that condition (4.10) is both necessary and sufficient as a condition for the existence of three-species equilibria.

Given that condition (4.10) is met, it remains to determine whether or not such three-species equilibria are asymptotically stable. [Evidently, if either parasitoid species goes extinct, the resultant two-dimensional system can be analysed using the results of Section 3.4.] Proceeding according to the standard methods for the analysis of a model's dynamics about equilibria, using equations (4.7) repeatedly, it is possible to find the following Jacobian matrix for the linearized flow about equilibria:

$$\begin{bmatrix} 1 & P^*f_1^{-1}df_1/dP|_{P^*} & Q^*f_2^{-1}df_2/dQ|_{Q^*} \\ 1 & -N^*df_1/dP|_{P^*} & 0 \\ 1 & P^*f_1^{-1}df_1/dP|_{P^*} & -N^*f_1df_2/dQ|_{Q^*} \end{bmatrix}$$

which we shall designate matrix (4.11). From Jacobian matrix (4.11), the eigenvalue equation can in turn be constructed, in polynomial form. Say that it has the following form:

$$\theta^3 - A\theta^2 + B\theta - C = 0$$

Asymptotic stability requires $|\theta_i| < 1$, for all

three θ_i roots. In three-dimensional systems,

like the present one, this condition is satisfied
when

$$|1+B| > |A+C| \quad \& \quad 1 - C^2 > |AC - B|$$

(May, 1974, p.220).

The properties of the local dynamics given by
(4.11) are discussed in some detail by May and
Hassell (1981). Given that both parasitoids attack
in a strongly clumped fashion, so that both k_i <<

1 , local asymptotic stability seems to be quite
likely. If one of the parasitoid species exhibits
no clumping, so that one $k_i > 1$, then it is still

possible to find asymptotically stable three-
species equilibria, but they are much less common.
Finally, while global asymptotic stability of
equilibria is not guaranteed by local asymptotic
stability, in that numerical examples with the
latter but without the former have been found, they
do seem to occur together if the model parameters
are well within the domain required for local
asymptotic stability (May and Hassell, 1981,
p.258).

4.2   ONE PREDATOR AND TWO PREY

Continuous-Time Models
The general continuous-time model of a simple
ecosystem with one predator and two prey is of the
form:

$$dx_1/dt = x_1 g_1(x_1, x_2, y) \qquad (4.12.1)$$

$$dx_2/dt = x_2 g_2(x_1, x_2, y) \qquad (4.12.2)$$

$$dy/dt = y g_3(x_1, x_2, y) \qquad (4.12.3)$$

where g_1 and g_2 are strictly decreasing

functions of y , while g_3 is a strictly

increasing function of both x_1 and x_2 . The

remaining features of the g_i functions can be

arbitrary: the two prey could be mutualists or competitors, providing only that one does not prey upon the other [a case dealt with in the next section], while any pattern of intraspecific density-dependence could be allowed.

From what is already known of the potential complexities of three-species ecosystems (e.g., Gilpin, 1975, 1979; Takeuchi and Adachi, 1983), it is clear that almost anything might happen in a system like (4.12). Achieving any concrete insights requires more explicit expressions for the g_i. This may be contrasted with two-species model ecosystems, about which fairly general insights were obtained from models with incompletely specified analogues of the present g_i functions, as discussed in Chapters Two and Three. Serious mathematical discussion of models based on specific forms of system (4.12) began with Cramer and May (1972), and further work has appeared by Fujii (1977), Vance (1978), and Takeuchi and Adachi (1983). Scientifically speaking, interest has revolved around the question of how similar the two prey species can be and yet be maintained in model ecosystems to which a predator species has been added, thereby focusing upon the issues of "competitive exclusion", "resource partitioning", "Gause's Law", species diversity, and so on.

A starting point for the analysis of model (4.12) is provided by the Lotka-Volterra case of strictly linear g_i functions:

$$dx_1/dt = x_1(\epsilon_1 - \alpha_{11}x_1 - \alpha_{12}x_2 - \alpha_{13}y) \qquad (4.13.1)$$

$$dx_2/dt = x_2(\epsilon_2 - \alpha_{21}x_1 - \alpha_{22}x_2 - \alpha_{23}y) \qquad (4.13.2)$$

$$dy/dt = y(-\epsilon_3 + \alpha_{31}x_1 + \alpha_{32}x_2 - \alpha_{33}y) \qquad (4.13.3)$$

where α_{13} , α_{23} , α_{31} , and α_{32} are all greater than zero. Let

$$A = \begin{bmatrix} \alpha_{11} & \alpha_{12} & \alpha_{13} \\ \alpha_{21} & \alpha_{22} & \alpha_{23} \\ -\alpha_{31} & -\alpha_{32} & \alpha_{33} \end{bmatrix} \qquad (4.14.1)$$

$$\underline{\varepsilon} = \begin{bmatrix} \varepsilon_1 \\ \varepsilon_2 \\ -\varepsilon_3 \end{bmatrix} \qquad (4.14.2)$$

and

$$\underline{x}^* = \begin{bmatrix} x_1{}^* \\ x_2{}^* \\ x_3{}^* \end{bmatrix} \qquad (4.14.3)$$

where $\underline{x}^*$ is such that system (4.13) is at equilibrium with all species present at the population densities that it specifies. Evidently, x^* need not exist in the first octant. From (4.13), $\underline{x}^*$ must be such that

$$\underline{\varepsilon} = A\underline{x}^* \qquad (4.15)$$

It is an elementary consequence of linear algebra that, if the determinant of A is not equal to zero, an $\underline{x}$ satisfying condition (4.15) exists and is unique, although not all of its elements need be strictly positive, so that it need not be an $\underline{x}^*$ which is in the model's relevant state-space. In any case, such potential three-species equilibria may be found from

$$\underline{x}^* = A^{-1}\underline{\varepsilon} \qquad (4.16)$$

where, by definition,

$$A^{-1}A = I$$

where I is the three-by-three identity matrix, having a principal diagonal of 1's and all other

elements zero. Given the definition of A^{-1}, the derivation of (4.16) from (4.15) requires only the

multiplication of both sides of (4.15) by A^{-1}. Given that one has found a suitable $\underline{x}^*$, with all elements strictly positive, its local stability properties are to be inferred from the Jacobian matrix of (4.13), in the usual fashion, this matrix being

$$\begin{bmatrix} -\alpha_{11}x_1^* & -\alpha_{12}x_1^* & -\alpha_{13}x_1^* \\ -\alpha_{21}x_2^* & -\alpha_{22}x_2^* & -\alpha_{23}x_2^* \\ \alpha_{31}y^* & \alpha_{32}y^* & -\alpha_{33}y^* \end{bmatrix} \qquad (4.17)$$

where this matrix is to be derived in the Exercises. This matrix in turn will lead to an eigenvalue polynomial of the form

$$\lambda^3 + a_1\lambda^2 + a_2\lambda + a_3 = 0 \qquad (4.18)$$

The explicit formulae for the roots of (4.18) are rather messy. However, as with the two-species case, it is possible to ascertain the basic properties of the real parts of the roots from relationships between the a_i's . In particular,

$\underline{x}^*$ is locally asymptotically stable if and only if
$$a_i > 0 \qquad (4.19.1)$$

for all three i , and
$$a_1 a_2 > a_3 \qquad (4.19.2)$$

This is one instance of the Routh-Hurwitz criteria for local asymptotic stability, where these criteria are also available for systems of higher dimension. [May (1974, p.196) gives all the practically useful Routh-Hurwitz criteria.]

Finally, if $\underline{x}^*$ is unstable and all the boundary points (origin, axes, and zero-faces) of the first octant are unstable, the three species populations must remain present, given an initial point inside the positive octant, but the density of at least one of them must fluctuate or grow without bound. [It may be possible to exclude unlimited growth of any species population density, given non-zero α_{ii} parameters for all three

species and the absence of mutualism.] Finding asymptotic trajectories for such systems analytically is generally not feasible, unless there are special symmetries. Therefore, numerical solutions will normally be of value in understanding the model dynamics.

Returning to the general model (4.12), similar lines of analysis can be followed with non-linear versions of this model, with the Jacobian matrix arising in the usual way, by partial

differentiation in the vicinity of equilibria. It
should be noted that equation (4.16) cannot be used
to find equilibria in such cases. However, the
Routh-Hurwitz criteria remain the same. Numerical
solutions will again be extremely helpful. More
sophisticated mathematical tools for analysis of
these higher-dimensional models are used in
Takeuchi and Adachi (1983); apparently there is a
wealth of opportunities for sophisticated applied
mathematics in the analysis of three-species
ecosystem models.

Discrete-Time Models: Polyphagous Parasitoids
Comins and Hassell (1976) initiated discussion of
the following sort of discrete-time, one-predator,
two-prey ecosystem:

$$x_{t+1} \quad = \quad x_t f_1(x_t, y_t, p_t) \qquad (4.20.1)$$

$$y_{t+1} \quad = \quad y_t f_2(x_t, y_t, p_t) \qquad (4.20.2)$$

$$p_{t+1} \quad = \quad x_t g_1(x_t, y_t, p_t)$$
$$\qquad \qquad + \; y_t g_2(x_t, y_t, p_t) \qquad (4.20.3)$$

with the f_i decreasing functions of p_t . Since

$p_{t+1} = 0$ if $x_t = y_t = 0$, this model is best

suited to parasitoid species and their host
species. Comins and Hassell (1976) specifically
considered functions of the form
$$f_i \quad = \quad \exp[r_i - c_i(x_t + \alpha y_t) - a_i p_t]$$
and
$$g_i \quad = \quad 1 - \exp(-a p_t)$$

Further discussion of such models is provided by
Hassell (1978, 1979).

Investigation of models like (4.20) is much
more difficult than that required for systems with
just one host, because the interaction between
competitors will not, normally, have the simple
all-or-none form that the host-parasitoid
interactions have. Thus it is very hard to treat
these models analytically, even with explicit f_i

and g_i . This is not to say that the only

available means of treating such models is
numerical solution. Rather, isocline equations for
the competitors can be found; equilibrium existence
can be assessed algebraically, graphically, or

numerically; Jacobian matrices and eigenvalue roots can be calculated; and parametric domains determining local equilibrium behaviour can be computed. Indeed, this is the approach of Comins and Hassell (1976). Once such things have been done, however, numerical solutions for discrete-time models will always be helpful, because there may be asymptotically stable attractors which are not revealed by local analysis.

4.3 THREE-SPECIES FOOD CHAINS

Continuous-Time Models
In a three-species food chain, a single prey species is subject to predation from a single species, the latter species in turn being subject to predation by a third species, a "top predator". The "bottom" prey species has no direct interaction with this top predator.

Mathematical analysis of three-species food chains began with Rescigno and Jones (1972) and Rosenzweig (1973). Further work on the problem is discussed by Freedman and Waltman (1977), who also provided a detailed treatment of a generalized model and some special cases. Here we will follow their analysis to illustrate the sort of work involved.

The model of interest is a straightforward extension of the predation model (3.25) discussed in Section 3.2:

$$dx/dt = xg(x) - yp(x) \qquad (4.21.1)$$
$$dy/dt = y[-r + cp(x)] - zq(y) \qquad (4.21.2)$$
$$dz/dt = z[-s + mq(y)] \qquad (4.21.3)$$

where r , s , c , and d are positive constants. As before, we make assumptions of the following kind:

all functions are smooth, differentiable, etc.;

$g(0) = \alpha > 0$; $dg(x)/dx < 0$ for $x \geq 0$; there exists a unique K such that $g(K) = 0$;

$p(0) = 0$; $dp(x)/dx > 0$ for $x \geq 0$; $q(0) = 0$; $dq(y)/dy > 0$ for $y \geq 0$.

Two equilibria which must exist are

$$E_1 = (x_1, y_1, z_1) = (0,0,0) \qquad (4.22.1)$$

and

$$E_2 = (K,0,0) \qquad (4.22.2)$$

Evidently, no equilibria of form $(a,0,b)$, $b > 0$,

are possible. On the other hand, equilibria with
$$E_3 = (x`,y`,0)$$

$x` > 0$ and $y` > 0$, are evidently possible, as
for model (3.25). Proceeding as for that model,
E_3 exists if and only if there is a solution

$x` > 0$ for
$$p(x`) = r/c \qquad (4.23.1)$$
such that
$$x` < K \qquad (4.23.2)$$
giving an equilibrium
$$E_3 = (x`,x`g[x`]/p[x`],0) \qquad (4.24)$$

Similarly, there may be an
$$E_4 = (x*,y*,z*) \qquad (4.25.1)$$

with all three population densities strictly
positive. As with conditions (4.23), the
conditions for the existence of this equilibrium
are that there be a $y*$ such that
$$q(y*) = s/m \qquad (4.25.2)$$
and
$$y* = x*g(x*)/p(x*) \qquad (4.25.3)$$
with $x* > 0$, and
$$z* = y*[-r + cp(x*)]/q(y*) \qquad (4.25.4)$$
such that $z* > 0$. Note that (4.25.3) implicitly
requires
$$-r + cp(x*) > 0$$
and thus
$$-r + cp(x*) > -r + cp(x`)$$
or
$$x* > x` \qquad (4.26.1)$$
when both arise, because $p(x)$ is a strictly
increasing positive-valued function of x . [This
is interesting biologically, because it shows
immediately that we must have higher equilibrium
prey population densities when there is a top
predator feeding on the "middle" species.] We also
must have $g(x*) > 0$ from (4.25.2), so that
analogously we must have
$$K > x* \qquad (4.26.2)$$
because of the nature of the $g(x)$ function.
Moreover, since
$$-r + cp(0) = -r < 0$$
if E_4 exists there must be an $x`$ such that

$$-r + cp(x`) = 0$$
because the continuity of $p(x)$ ensures the
continuity of the function
$$s(x) = r + cp(x)$$

while $s(0) < 0$ and $s(x*) > 0$, necessitating the existence of some $x^`$ in the interval $(0, x*)$ for which $s(x^`) = 0$. Condition (4.26) also guarantees that $K > x^`$, when E_4 exists, so that we can

conclude that E_3 exists whenever E_4 exists.

The Jacobian matrix, say J , giving the linear flow about any particular point in the first octant is given by

$$
\begin{bmatrix}
xdg/dx+g(x)-ydp/dx & -p(x) & 0 \\
cydp/dx & -r+cp(x)-zdq/dy & -q(y) \\
0 & mzdq/dy & -s+mq(y)
\end{bmatrix} \quad (4.27)
$$

Evidently

$$
J(E_1) = \begin{bmatrix}
\alpha & 0 & 0 \\
0 & -r & 0 \\
0 & 0 & -s
\end{bmatrix}
$$

which necessarily gives rise to local instability about the origin. Likewise

$$
J(E_2) = \begin{bmatrix}
Kdg(K)/dx & -p(K) & 0 \\
0 & -r+cp(K) & 0 \\
0 & 0 & -s
\end{bmatrix}
$$

with eigenvalues given by the principal diagonal. Evidently $-s$ and $Kdg(K)/dx$ are strictly negative eigenvalues. Thus local asymptotic stability for equilibrium E_2 requires only

$-r + cp(K) < 0$

as for the analogous equilibrium of model (3.25). Thus, if

$cp(K) > r$

E_2 is unstable, and all boundary points along the

x-axis are unstable. All of these results are exactly the same as those found for the x-axis in the analysis of model (3.25).

As was discussed for predator-prey model (3.25), when conditions (4.23) are met, the system behaviour on the (x,y) plane is asymptotic approach to equilibrium E_3 or limit cycles about it.

Behaviour near the (x,y) plane thus depends on the local (x,y,z) dynamics about E_3 and the (x,y,z) dynamics near limit cycles on the (x,y) plane. The analysis of local dynamics about E_3 depends on the Jacobian matrix

$$\begin{bmatrix} x\,dg/dx+g(x\,)-y\,dp/dx & -p(x\,) & 0 \\ cy\,dp/dx & 0 & -q(y\,) \\ 0 & 0 & -s+mq(y\,) \end{bmatrix} \quad (4.28)$$

Thus the local population dynamics of z are given by

$$dz/dt = z[-s + mq(y\,)]$$

near E_3. Therefore, if $mq(y\,) > s$, E_3 is locally unstable, and a trajectory beginning in some local neighbourhood of E_3 moves away from it and the (x,y) plane. Note that $s > mq(y\,)$ does not guarantee the local asymptotic stability of E_3 because it could be unstable with respect to x and y perturbations on the (x,y) plane. In any case, if (i) there are no non-trivial, stable, periodic solutions in the (x,y) plane, (ii) cp(K) > r , and (iii) $mq(y\,) > s$, then all three species will persist, given arbitrary positive initial densities (vid. Theorem 3.1, Freedman and Waltman, 1977a).

When there are non-trivial, stable, periodic trajectories on the (x,y) plane, say

$$(x,y) = [\phi(t),\phi(t)]$$

then the analogous condition (iii) for persistence of all three species must include

$$m \int_o^T q[\phi(t)]/T \, dt > s$$

where the integral is a precise analogue of $q(y\,)$ (Theorem 3.2, Freedman and Waltman, 1977a). This second result in effect averages the local dynamics of z over the range of its prey density fluctuations.

The behaviour away from the boundaries of the

first octant of (x,y,z) space is discussed in detail by Freedman and Waltman (1977a). They show that the trajectories are always bounded, with none of the species population densities becoming arbitrarily large as t becomes arbitrarily large. They also provide qualitative conditions on the terms of $J(E_4)$ for asymptotic stability of E_4.

In particular, like the results of the analysis of model (3.25), they show that: (i) if the slope of $x^*g(x^*)/p(x^*)$ is strictly positive, then E_4 is unstable; and (ii) if the slope of $x^*g(x^*)/p(x^*)$ is strictly negative and the slope of $y^*/q(y^*)$ is zero or negative, then E_4 is locally asymptotically stable. The questions of limit cycle and strange attractor occurrence remain unexamined, formally. However, the possibility of limit cycles in model (3.25) ensures that they can arise in model (4.21), if only by making the latter approach the former arbitrarily closely in cases where limit cycles arise in the two-species model.

Discrete-Time Models: Parasitoid-Hyperparasitoid Systems

The first published analysis of a three-species, discrete-time, food-chain model was that of Beddington and Hammond (1977). They combined density-dependent host reproduction with the Nicholson-Bailey parasitoid attack equation, here given as equation (3.46). Other models of this type are discussed by Hassell (1978, pp.157-163; 1979) and, in greatest detail, by May and Hassell (1981). Here we will consider a model treated in May and Hassell (1981):

$$N_{t+1} = \lambda N_t f_1(P_t) \tag{4.29.1}$$

$$P_{t+1} = N_t[1-f_1(P_t)]f_2(Q_t) \tag{4.29.2}$$

$$Q_{t+1} = N_t[1-f_1(P_t)][1-f_2(Q_t)] \tag{4.29.3}$$

with all variables and functions as before in Sections 3.4 and 4.1.
At equilibrium we must have

$$1 = \lambda f_1(P^*) \tag{4.30.1}$$

from (4.29.1). Equations (4.29.2) and (4.29.3) give us

$$Q^* = [1 - f_2(Q^*)]P^*/f_2(Q^*)$$

or

$$P^* = Q^*f_2(Q^*)/[1 - f_2(Q^*)] \qquad (4.30.2)$$

and

$$N^* = P^* + [1 - f_1(P^*)]f_2(Q^*)$$

$$= P^* + (1 - \lambda^{-1})f_2(Q^*)$$

$$= \lambda P^*/[(\lambda-1)f_2(Q^*)] \qquad (4.30.3)$$

It is worth recalling that in the host-parasitoid analogue of model (4.29), given by (3.38), equation (3.39.2) yields

$$N^* = \lambda P^*/(\lambda-1)$$

Evidently, we must have $f_2(Q^*) < 1$ in any

three-species equilibrium. [Otherwise $Q^* = 0$, from (4.29.3).] Thus equilibrium-point equation (4.30.3) shows that hyperparasitoids result in increased "bottom" host population densities, at equilibrium. [Precisely the same effect was demonstrated for the analogous continuous-time model, in result (4.26.1).] On the other hand, it is interesting to note that P^* has the same value in either two- or three-species food chain cases.

Proceeding as for the competing parasitoid model, with f_1 and f_2 conforming to the form of

equations (4.6), three-species equilibrium existence requires

$$a_1/a_2 < k_1(\lambda^{1/k_1} - 1) \qquad (4.31)$$

so that the hyperparasitoid can be maintained at equilibrium only if a_2, its searching efficiency, is sufficiently high.

The stability criteria for equilibrium (4.30) when the model has the parasitism probabilities defined by searching efficiency equations of form (4.6) are quite complex (May and Hassell, 1981, p.257), and their derivation would not be very useful written out here. As is reasonable from the earlier results for parasitoid systems of this kind, these conditions show that stability is facilitated by small k_i and large a_2. That is

to say, parasitoid and hyperparasitoid attack must be highly clumped together with extremely efficient hyperparasitoid searching.

4.4 EXERCISES

Elementary

1. Derive (4.8).

2. Derive (4.9.2).

3. Derive (4.17).

4. Derive (4.27).

Intermediate

5. Derive (4.31).

6. Analyse:
$$dx_1/dt = x_1(\epsilon - 2x_1 - 1.75x_2 - dy)$$
$$dx_2/dt = x_2(\epsilon - 2x_2 - 1.25x_1 - dy)$$
$$dy/dt = y(-1 + bx_1 + bx_2)$$

Advanced

7. Develop and analyse a three-species model with one predator, its prey, and a predation-free competitor of the prey species, where the density-dependent effects of one species on another are not linear.

Chapter Five

COMPLEX ECOSYSTEMS

The three-species "simple ecosystem" models of the
preceding chapter teetered on the edge of
analytical intractability. In some cases, very
little useful analysis was possible, at least with
any generality, as illustrated by the discussion of
the models of Section 4.2. In other cases, fairly
general analysis was possible, with some useful
insights. Some of the findings of Section 4.3 are
examples on this side of the question. When one
proceeds to four-species models of any generality,
insightful analysis becomes extremely difficult.
Thus we instead proceed to a discussion of
ecosystem models with an arbitrary number of
species, therefore called "complex". The few
general principles of analysis will be discussed,
and questions of possible empirical significance
raised. As will become clear, most of the basic
ideas of this chapter have already come forward in
previous chapters. Indeed, it is doubtful that we
would have much intuition for the possibilities
arising in complex ecosystem models were it not for
what has been discovered in far simpler models.

5.1 LOCAL EQUILIBRIUM STABILITY

Once an ecosystem model has been developed, one of
the most important avenues for mathematical
analysis is the discovery of equilibria and the
evaluation of the local dynamics about such
equilibria. Normally, the latter can be done by
obtaining the eigenvalues of the Jacobian matrix of
first partial derivatives of the system's dynamical
equations at the equilibrium point, as we have now
seen a number of times. Surprisingly, it turns out
that there are some general scientific issues which
arise from the study of such Jacobian matrices
derived from complex ecosystem models.

Time-Structure and Local Asymptotic Stability

For continuous-time systems, the necessary and sufficient condition for local asymptotic stability is that the real parts of all Jacobian matrix eigenvalues be strictly negative. If some real parts are zero, while the rest are strictly negative, then the equilibrium is merely "stable" or "neutrally stable", locally. [Scientifically, there are good reasons for doubting that such merely stable equilibria are of material significance, as illustrated in our discussion of the classical Lotka-Volterra predator-prey model.] Otherwise, the equilibrium is unstable. While these eigenvalues might be found explicitly in some instances, such as two-species models, generally asymptotic stability is best ascertained indirectly using the Routh-Hurwitz criteria or some other matrix eigenvalue theorem (May, 1974, Appendix II; Pullman, 1976, Chapter IV).

As we have seen, for discrete-time systems the condition for local asymptotic stability of equilibria is that the absolute value, or modulus, of all system eigenvalues be less than one. Otherwise the system is merely stable or, if any absolute values are strictly greater than one, unstable. This condition is more severe than that for continuous-time models, reflecting the destabilizing effects of strong negative feedback in discrete-time systems. In effect, such models are subject to "over-shoot destabilization".

Though we have illustrated the use of time-lags only in single-species discrete-time models, they can be introduced in all types of model, discrete-time and continuous-time, with any number of species (Hutchinson, 1948; Wangersky and Cunningham, 1956, 1957a,b; May, 1973, 1974; Wangersky, 1978). These models tend to show that time-lags destabilize equilibria, in that the parametric conditions on local asymptotic stability become more stringent. This is analogous to the contrast between discrete-time and continuous-time models.

Perhaps one of the very few general conclusions the study of theoretical ecology has to offer is that ecosystems subject to large, discrete-time, lag effects should be less stable in the sense of local asymptotic stability of population density equilibria. [This conclusion need not hold up when there are some kinds of "distributed" time-lags, which actually smooth out tendencies to divergent oscillation (Cushing, 1977; MacDonald, 1978).] However, the empirical difficulty in testing this conclusion is how the degree of time-lag in any particular ecological interaction is to be determined. In some cases, this problem is not difficult: univoltine parasitoid-host complexes are subject to an annual time-lag. Still more crudely, it might be argued that ecosystems with short annual growing seasons are subject to a higher degree of time-lag than those with continuous macrophytic vegetative growth, and compare stability properties accordingly. It is not my intention to defend any of these indirect avenues of evaluation in particular, only to suggest one of the few likely avenues of empirical investigation of what may be a well-founded theoretical conclusion.

Arbitrary Complexity and Local Stability

Thanks to Robert M. May, one of the central themes in theoretical ecology has been the question of the effects of increasing complexity on the local stability of model ecosystem equilibria. May's work in turn was inspired by the widespread view that "increased trophic web complexity leads to increased community stability" (May, 1971, p.59).

May's first counter-argument to the view that this assertion is true <u>for strictly mathematical reasons</u> was a comparison of classical, two-species, Lotka-Volterra, predator-prey, model ecosystems with n-predator, n-prey, Lotka-Volterra, model ecosystems. As we saw in Chapter Three, in the absence of intraspecific density-dependent effects on population growth, the former model exhibits strictly neutral stability of equilibria. The analogous n-predator, n-prey, Lotka-Volterra model is given by May (1971) as follows:

$$dx_j/dt = x_j(a_j - \Sigma_k \alpha_{jk} y_k) \qquad (5.1.1.j)$$

$$dy_j/dt = y_j(-b_j + \Sigma_k \beta_{jk} x_k) \qquad (5.1.2.j)$$

with the range of k in the summations being from 1 to n, while both types of equation appear n times for the n distinct values of j. The local stability properties of this model depend upon a $2n \times 2n$ Jacobian matrix with the following form where the equilibrium values of the x_j and y_j are indicated by x_j^* and y_j^*, respectively.

$$\begin{bmatrix}
0 & \cdots & \cdots & 0 & -\alpha_{11}y_1^* & \cdots & \cdots & -\alpha_{11}y_{11}^* \\
\vdots & & & & & & & \vdots \\
 & & & & & -\alpha_{jj}y_j^* & & \\
\vdots & & & & & & & \vdots \\
0 & \cdots & \cdots & 0 & -\alpha_{n1}y_1^* & \cdots & \cdots & -\alpha_{nn}y_n^* \\
\beta_{11}x_1^* & \cdots & \cdots & \beta_{1n}x_n^* & 0 & \cdots & \cdots & 0 \\
\vdots & & & & & & & \vdots \\
\vdots & & & & & & & \vdots \\
\beta_{n1}x_1^* & \cdots & \cdots & \beta_{nn}x_n^* & 0 & \cdots & \cdots & 0
\end{bmatrix}$$

This matrix will be designated (5.2). It arises because: (a) $\partial[dx_j/dt]/\partial x_k = 0$, for k not

equal to j , while likewise $\partial[dy_j/dt]/\partial y_k = 0$;

and (b) $\partial[dx_j/dt]/\partial x_j = a_j - \Sigma_k \alpha_{jk}y_k = 0$

at equilibrium, and likewise $\partial[dy_j/dt]/\partial y_j = 0$.

The eigenvalues of $2n \times 2n$ matrices of form (5.2) occur in n pairs of the form
$$\lambda_m = \pm (x_m + iy_m) \tag{5.3}$$

[This result is a basic theorem of matrix algebra.] Thus, either all the x_m's are zero and the system

is merely stable, or it is unstable. Though the one-predator, one-prey version of (5.1) always has $x_m = 0$, this is true in the n-predator, n-prey

case only if special symmetries arise, symmetries which are biologically implausible in the extreme (May, 1974, pp.50-53).

An analogous result may be proven for systems more general than (5.1) (May, 1971, p.76). Take the system made up of N equations of the form
$$dx_j/dt = f_j(x_j)g_j(\{x_{k,k\neq j}\}) \tag{5.4}$$

where the set notation, $\{\}$, indicates a set of variables of the indicated type. We also take $f_j(x_j) = 0$ only at $x_j = 0$. That is, we assume

that there are no carrying capacities $K_j > 0$ such

that $f_j(K_j) = 0$, and that g_j does not depend on x_j . At equilibria $\underline{x}^*$ with all $x_j > 0$, we must have $g_j = 0$. Therefore,

$$\partial[dx_j/dt]/\partial x_j = g_j(\underline{x}^*)\partial f_j/\partial x_j + f_j(x_j^*)\partial g_j/\partial x_j$$
$$= g_j(\underline{x}^*)\partial f_j/\partial x_j = 0 \qquad (5.5)$$

for all dx_j/dt . Thus the Jacobian for $\underline{x}^*$ of model (5.4) must have strictly zero terms on the principal diagonal. It is a basic result of matrix theory that the trace of a matrix

$$Tr(M) = \Sigma_j \, a_{jj} = \Sigma_j \, \lambda_j \qquad (5.6)$$

where all summation proceeds from 1 to n and where the λ_j are the eigenvalues of the matrix M (May, 1974, p.195). This sum will always be a real number for real matrices, because the complex eigenvalues of real matrices occur in conjugate pairs, with an eigenvalue equalling $x - iy$ for every $x + iy$ eigenvalue. From (5.5) and (5.6), we can conclude that all the real parts of the eigenvalues of system (5.4) will be zero, or the system will be unstable. In either case, the model cannot have interior equilibria which exhibit local asymptotic stability. This generalizes the conclusion obtained for model (5.1).

For ecosystems lacking the special structural constraints of model (5.4), analytical results of this kind are not so readily obtained. Instead, research on the stability properties of local equilibria in general types of ecosystem models has used arbitrary matrices generated pseudo-randomly by "Monte Carlo" computer programs (e.g., Gardner and Ashby, 1970; McMurtie, 1975) or limiting asymptotic approximations to the statistical properties of such matrices (e.g., May, 1972; May, 1974, pp.64-68). For such arbitrary matrices, the likelihood of all eigenvalues having strictly negative real parts decreases with increasing species number and increasing interaction between species. Thus arbitrary "ecosystem models" of increased complexity have equilibria which are less likely to exhibit local asymptotic stability, since the Jacobian matrices associated with their equilibria will also be arbitrary. Thus there is indeed no strictly mathematical reason to expect

increased ecosystem complexity to be associated
with increased ecosystem "stability", in the
specific sense of local asymptotic stability of
equilibria.

Ecosystem Model Structure and Local Stability

Theoretical research on the question of ecosystem
stability and complexity has thus turned to a
different question: what sort of ecosystem model
structure is required to make complex ecosystem
models as likely to be stable as simple ecosystem
models, if not more likely?

Again, this work has relied upon Monte Carlo
numerical exploration and analytical
approximations, as opposed to exact mathematical
analysis (e.g., De Angelis, 1975). One of the
clearest results of this research is that it is
indeed possible to find realistic rules of
construction which allow local asymptotic stability
of equilibria of complex ecosystem models (May,
1979). One example of such rules is that "the
higher trophic level species experience a strong
self-damping force which controls their population
growth" (De Angelis, 1975, p.242). This seems like
a fairly reasonable biological constraint, which
could easily be met by species which exhibit
territoriality, allelopathy, and so on.

Even if a theoretical result seems valid, in a
formal sense, it may not hold up empirically. One
quite general corollary of the mathematical results
is that increased compartmentalization of
ecosystems should facilitate stability, since
species interactions are thereby reduced. Yet there
is no evidence that ecosystems are organized in
this fashion (Pimm and Lawton, 1980). Many
obscurities remain.

5.2 GLOBAL COMPLEX ECOSYSTEM DYNAMICS

As was illustrated in earlier chapters, ecological
models may have several equilibria, more than one
of which may be stable. Moreover, some ecological
models may exhibit dynamics in which limit cycles
occur, particularly those involving predation, and
these limit cycles too need not be unique.
Finally, there are the elusive mysteries of strange
attractors, seen briefly in the numerical
exploration of the discrete-time logistic requested
in the Exercises at the end of Chapter One.

Therefore, any view of the dynamics of complex

ecosystems which is entirely centered upon the local stability of a single equilibrium is, as a basis for generalization, radically deficient. What is necessary is characterization of the global dynamical properties of the model ecosystem throughout its admissible state-space. The need for such characterization is the unifying theme of the "resilience" debate launched by Holling (1973), and since taken up by many others (e.g., Beddington et al., 1976; May, 1977). Evidently, if the separatrices about a locally asymptotically stable equilibrium are such that a reasonably small, though greater than infinitesimal, perturbation can carry the system beyond these separatrices, then that equilibrium is stable but not resilient. To assess such equilibrium properties, and their analogues for limit cycles and strange attractors, all attractor trajectories and their basins of attraction must be obtained.

Methods of finding attractor trajectories have already been discussed. Evidently, the first step is that of finding equilibria and assessing their local stability. Next the existence of oscillatory trajectories about equilibria should be examined. The Poincaré-Bendixson Theorem has already been mentioned as one device for testing for such oscillatory trajectories, in planar ecological models (cf. Hirsch and Smale, 1974). Another useful result is the Hopf Bifurcation Theorem (Marsden and McCracken, 1976). This theorem is concerned with the conditions under which an unstable equilibrium will be surrounded by a limit cycle in continuous-time models, given that the stability of the equilibrium can be shown to depend upon a "tunable" parameter having specific properties. Finally, there are results such as those of Li and Yorke (1975) concerning the existence of strange attractors or "chaos".

Given that one has found the attractor trajectories of interest, their domains, or basins, of attraction remain to be considered. [That is, over what part of the admissible state-space of the system is asymptotic convergence to a particular attractor trajectory assured?] Here the most useful tools are Liapunov and Liapunov-like functions, which may be used in both continuous-time and discrete-time systems (Goh, 1980, passim). With these tools it is possible to show global convergence to a strange attractor in a discrete-time system, for example (Goh, 1980, pp.114-115). Thus very powerful tools are available, in principle. That is, given that we are sufficiently ingenious in our discovery of attractor trajectories and Liapunov functions, analysis can uncover a great deal. The problem is, however, that few of these tools of global analysis can be handled in the same "standard recipe" fashion that we follow in examining local equilibrium stability. Few of the applications of these techniques will prove of much value in the analysis of ecosystems with more than three species.

One exception to this practical rule is Goh's (1977) study of global stability conditions for unique asymptotically stable equilibria, elaborated upon and discussed further in Case and Casten (1979) and Goh (1980, Ch.5). Goh's most important result is the following theorem:

Given: (i) A Lotka-Volterra model for m species,

$$dx_i/dt = x_i(b_i + \Sigma_j a_{ij}x_j) \qquad (5.7.i)$$

where i = 1, . . , m and the limits of summation are 1 and m , with a feasible equilibrium $\underline{x}*$ at which $x_i* > 0$ for all i ;

and (ii) a constant positive diagonal matrix C

such that $CA + A^T C$ is negative definite (see below), where A is the matrix of the a_{ij}'s from

equations (5.7.i) and T indicates matrix transposition. [Matrix transposition involves reversing the position of matrix elements above and

below the principal diagonal so that $a_{ij}^T = a_{ji}$.]

Then: Equilibrium $\underline{x}*$ is globally stable within the admissible state-space.

The proof of this theorem is given in Goh (1977). It makes use of the Liapunov function defined by

$$V = \Sigma_i c_i(x_i - x_i* - x_i*\ln[x_i/x_i*]) \qquad (5.8)$$

where the c_i are the c_{ii} elements of the matrix

C . [This is really a generalization of the Liapunov function that we discussed in Section 3.1.] In the proof, it is shown that

$$2dV/dt = (\underline{x}-\underline{x}*)^T(CA + A^T C)(\underline{x}-\underline{x}*) \qquad (5.9)$$

Here is where the negative definite matrix requirement comes in. If a matrix M is negative

definite, then $\underline{v}^T M \underline{v} < 0$, for all $\underline{v}$, by
definition. Thus the stipulation that there be a

C such that $CA + A^T C$ is negative definite is
equivalent to the demand that dV/dt be strictly
less than zero.

Goh (1977) provides two other interesting
results. Firstly, it is shown that, for
Lotka-Volterra matrices constructed in accordance
with Levins' (1968, pp.50-55) recipe for a
competitive community matrix, local asymptotic
stability implies global asymptotic stability.
Secondly, Goh generalizes the Lotka-Volterra
version of the theorem to nonlinear ecosystem
models of form

$$dx_i/dt = x_i F_i(x_1 . . x_m) \qquad (5.10.i)$$

where $i = 1, . . . , m$ and $CA(\underline{x}') + A^T(\underline{x}')C$

replacing $CA + A^T C$ in the theorem, where $A(\underline{x}')$
is the matrix of first partial derivatives of the
F_i at $\underline{x}'$ and $\underline{x}'$ may be any point in the

admissible state-space. In effect the new A
matrix is variable over the whole state-space,
whereas before, in the Lotka-Volterra case, it was
constant, given particular parameter values. Thus
the condition on $A(\underline{x}')$ must be met everywhere, an
extremely demanding requirement, and one which
would be practically quite difficult to assess for
models of moderate complexity.

Even in the Lotka-Volterra case, there is no
guarantee that an appropriate C will be easy to
find, even when an equilibrium is in fact globally
asymptotically stable. And in the generalized case
of model (5.10.i) it will usually be impossible to
use Goh's theorem. None the less, these results do
constitute a first step in the exploration of the
essentially unknown realm of global complex
ecosystem dynamics.

Chapter Six

MIGRATION

To this point, we have assumed that all modelled populations have been growing, competing, and so on in splendid isolation from neighbouring ecosystems. But of course most populations receive frequent migrants from other ecosystems, migrants which may be critical in retrieving a small population from extinction, or may have essentially negligible effects on a rapidly growing population. The determination of the material importance of migration for ecological dynamics is naturally a theoretical problem of fundamental importance. Here we shall address this problem in terms of the categories given by each of the topics taken on to this point: population growth, competition, predation, and ecosystem models.

In addition to this division of our subject matter, we will be treating three different possible patterns of migration. The first is that of "recipient peripheral populations". It is assumed that such migration proceeds by the addition of a constant number of individuals per unit of time, these individuals deriving from a very large population of stable density inhabiting a nearby ecosystem. It is implicitly assumed that the dynamics of the explicitly modelled population(s) have no effect on the number of migrants received per unit time. The second possible pattern of migration to be discussed is that of "migrant pools". Migration models of this type assume that populations contribute emigrants to a pool of migrants as well as receiving immigrants from this pool. It is also assumed that the net flux between modelled populations and the migration pool is a function of the density of modelled populations, the notion being that migrants are dispersing from areas of higher

intraspecific competition or low prey availability
to areas of reduced competition or increased prey
densities, respectively. The third pattern which
will be discussed is migration between "two
habitats", where migration is a simple linear
function of emigrant population density in the
initial habitat. This last pattern is the most
common pattern of migration assumed in ecological
models, with the most elaborate theoretical
development.

6.1 POPULATION GROWTH WITH MIGRATION

Recipient Peripheral Populations
The basic idea for the recipient peripheral
population model of migration comes from Holt
(1983a,b), who has discussed both continuous-time
and discrete-time versions of this model. Our
development of this model follows his in a general
way.
   We begin with a continuous-time analysis. Let
x be the population density of a peripheral
population receiving migrants from a larger
population whose density it does not affect. We
let c represent the rate of immigration into the
peripheral population, where c must be strictly
positive. For a population growing on its own
without time-dependent characteristics or
time-lags, we can immediately write

$$dx/dt = xg(x) + c \qquad (6.1)$$

where we take $g(x)$ to be continuous and
differentiable. All equilibria arise when we have
x^* such that

$$g(x^*) = -c/x^* \qquad (6.2)$$

The local asymptotic stability of such equilibria
depends upon

$$
\begin{aligned}
d\{dx/dt\}/dx &= d\{xg(x) + c\}/dx \\
&= g(x) + xdg/dx \\
&= -c/x + xdg/dx \qquad (6.3)
\end{aligned}
$$

evaluated at x^* satisfying equation (6.2), since
this is the term of largest magnitude in the Taylor
expansion giving the population dynamics in the
immediate vicinity of x^* . Evidently, we require

$$-c/x^* + x^*dg(x^*)/dx < 0$$

or

$$dg(x^*)/dx < c/x^{*2} \qquad (6.4)$$

Evidently, this condition is more readily met than
the comparable condition in the absence of
migration,

$$dg(x^*)/dx < 0$$

since both c and x^* must be greater than zero.

Now let us consider a somewhat less general
set of models, those for which there is a K > 0
such that g(K) = 0 and dg/dx < 0 for all x .
It is at once obvious that, for any x* meeting
condition (6.2), we also have local asymptotic
stability. Given the nature of g(x) , we must
have x* > K in order that g(x*) = -c/x* , since
g(K) = 0 . Therefore, immigration raises the
equilibrium level of the population density, a
result which is not particularly surprising. In
the Exercises, you are to construct an argument
showing that local asymptotic stability also gives
rise to global asymptotic stability for population
growth models of this particular type.

To return to the general population growth
model, there is also the possibility of g(x)
functions which do not simply decrease with
increasing values of x . Some of the consequences
of this can be seen from graphs of g(x) and the
function -c/x against x , as in Figure 6.1.
Evidently, at intersections of these two functions,
we have equilibria for population growth with
migration, although these equilibria may be stable
or unstable. Evaluating the stability of the
equilibria actually depends on a composite function

$$m(x) = g(x) + c/x \qquad (6.5)$$

since we can write (6.1) as

$$dx/dt = xg(x) + c = x[g(x) + c/x] = xm(x)$$

Similarly, local asymptotic stability criterion
(6.4) can also be seen as equivalent to

$$dm(x*)/dx < 0$$

since this expression can be re-arranged as

$$dg/dx - c/x^2 < 0$$

or

$$dg/dx < c/x^2$$

as in (6.4). Thus local asymptotic stability turns
on whether or not m(x) is decreasing at x* . One
point which should be made here is that plots of
m(x) against x will have the same generic
features as the plots of analogous g(x) functions
in models without migration. Thus they will be
either above or below the x-axis, generically
crossing at single points. When the m(x)
function crosses from above, an equilibrium will be
locally asymptotically stable. When it crosses
from below, the equilibrium will be unstable,
defining a threshold between the domains of
attraction of two alternative asymptotic
trajectories. However, an important difference is
that the number of equilibria with migration may be

different from that without migration.

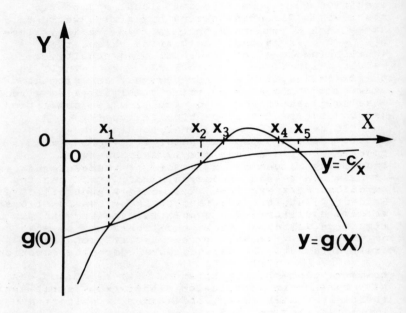

Figure 6.1. Plot of $g(x)$ and $-c/x$ functions for a model of population growth with immigration of the form of (6.1).

We do not need to plot such $m(x)$ functions explicitly, however. We can get all the required information from figures like Figure 6.1. When $g(x)$ is above $-c/x$, we have $m(x) > 0$, and vice versa. However, it is not the case that when $g(x)$ crosses $-c/x$ from below that the resulting equilibrium is necessarily unstable. The equilibrium at x_1 in Figure 6.1 is in fact locally

asymptotically stable. From the graph, we
evidently have $g(0) + c/0 > 0$, so that $m(0) > 0$.
Likewise, for x in the interval (x_1, x_2) , we
have $g(x) < -c/x$, so that
    $g(x) + c/x < 0$
or $m(x) < 0$. Thus, for x in the interval
$(0, x_2)$, the population growth trajectory converges

on x_1 . Biologically, what appears to be occurring

is the stabilization by immigration of a population
which cannot grow on its own at such low densities.
With the removal of migration, it is apparent that
for x in the interval $(0, x_3)$, population density

asymptotically falls to zero. In any event, it is
interesting that migration can result in stable
population density equilibria at densities where
the $g(x)$ function is increasing. The additional
equilibrium at x_5 must be locally asymptotically

stable, since there $g(x)$ is decreasing. In this
region of the state-space, migration has only
increased the equilibrium population density a bit,
from x_4 to x_5 . Over the entire state-space,

however, it should be apparent that migration can
give rise to new population density equilibria and
radically alter the asymptotic results of
population growth.
    One additional possibility is the elimination
of population growth equilibria as a result of
migration. This is shown in Figure 6.2.
Evidently, the effect here is one of migration
removing lower population density equilibria,
leading to the achievement of much higher
asymptotic population density levels. Finally, it
should be noted that as x increases to ∞ , the
function c/x converges to zero, so that at very
high population densities immigration has little
effect on population growth, in accordance with
intuition.
    Let us consider a specific example, with
    $g(x) = r(1 - x/K)$ (6.6)
our now banal logistic equation for the net
reproductive rate of individuals. Equilibria arise
when
    $-c = rx(1 - x/K)$
or
    $x^2 - Kx - Kc/r = 0$
This is a quadratic polynomial with roots given by

x* in
$$2x^*_\pm = K \pm \sqrt{K^2 + 4Kc/r} \qquad (6.7)$$

Since K , c , and r are all strictly positive,
both roots are real numbers. In addition, since
the quantity under the square root is greater than

K^2 , one of these roots has x* < 0, making it
inadmissible, while the other has x* > K . [This
second result is of course in accord with our
earlier general analysis for models with decreasing
g(x) functions.] In this case, we also have

$$dg(x^*)/dx = -r/K < 0 < c/x^{*2}$$
ensuring asymptotic stability of the equilibrium.

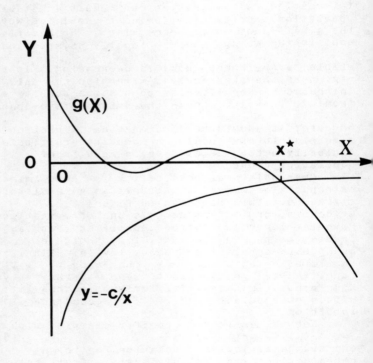

Figure 6.2. Plot of g(x) and -c/x
functions for population growth with
immigration as in model (6.1).

Now let us consider the discrete-time analogue of model (6.1):

$$x(t+1) = R[x(t)]x(t) + c \qquad (6.8)$$

Evidently, there are equilibria at

$$(R[x^*] - 1)x^* + c = 0 \qquad (6.9)$$

or

$$x^* = c/\{1 - R[x^*]\} \qquad (6.10)$$

We proceed to analyse the stability properties of such equilibria using a Taylor expansion for R about x^*. Let

$$x(t) = x^* + \varepsilon(t)$$

Then we have

$$x^*+\varepsilon(t+1) = R[x^* + \varepsilon(t)]\{x^*+\varepsilon(t)\} + c$$
$$\approx \{R[x^*] + \varepsilon(t)dR^*/dx\}\{x^*+\varepsilon(t)\} + c$$

$$= R^*x^* + c + R^*\varepsilon(t)$$
$$+ \varepsilon(t)x^*dR^*/dx + O(\varepsilon^2)$$

Using (6.9) twice, we find

$$\varepsilon(t+1) \approx \varepsilon(t)\{1 - c/x^* + x^*dR^*/dx\}$$

and thus

$$\Delta\varepsilon \approx \varepsilon\{x^*dR^*/dx - c/x^*\} \qquad (6.11)$$

[This result may be compared with the analogous one for population growth in discrete-time without migration, equation (1.64). It is also somewhat analogous to model (6.1).] Local asymptotic stability of equilibrium x^* therefore requires

$$b = x^*dR^*/dx - c/x^* \qquad (6.12)$$

in the interval $(0, -2)$ for the same reasons as those discussed in Section 1.3; if b is too large in magnitude, then there is oscillatory over-shoot of the equilibrium, giving rise to limit cycles, strange attractors, or departure to the basin of attraction of another population trajectory.

Again, for concreteness let us consider the example afforded by the discrete-time logistic model, with

$$\Delta x = rx(1 - x/K) + c$$

and thus

$$x(t+1) = [1+r - rx(t)/K]x(t) + c \qquad (6.13)$$

As before, we require $x^* > 0$ as given by equation (6.7), where $x^* > K$. In order to investigate the value of b, we substitute $dR^*/dx = -r/K$ into (6.12), obtaining

$$b = -rx^*/K - c/x^*$$

Since all parameters on the RHS are strictly positive, we must have $b < 0$. The sole condition on local asymptotic stability of x^* is then

$$2 > rx^*/K + c/x^* \qquad (6.14)$$
$$> r + c/x^* > r$$

This may be contrasted with the condition on the local asymptotic stability of K in the discrete-time logistic model without immigration, for which $2 > r$ was sufficient with strictly positive r and K values. Thus immigration destabilizes such equilibria by increasing the tendency to oscillatory over-shoot. However, this is by no means a necessary result of the effect of immigration on population growth in discrete-time models. From (6.12), it is apparent that immigration might stabilize equilibria when $dR*/dx$ is greater than zero, though only at low values of $x*$. For discrete-time models, as for continuous-time, the larger the value of $x*$, the less the effect of migration on the local asymptotic stability of equilibria.

Migrant Pool Model

Now let us consider population growth with migration conforming to a "migrant pool" assumption. Here we will be assuming that migration will tend to increase the density of low density populations and decrease the density of high density populations. Explicitly, this assumption may be represented by

$$dx/dt = xg(x) + xm(x)$$
$$= x[g(x) + m(x)] \qquad (6.15)$$

where we take

$$g(0) > 0 \ ; \quad m(0) > 0 \ ;$$
$$dm/dx < 0 \ ; \quad K \text{ such}$$
$$\text{that } g(K) = 0 \ , \quad K > 0 \ .$$

In the absence of further specification of g, it is difficult to say more about the population dynamics which can arise from (6.15). In effect, we have simply defined a new population growth function, $g + m$, which may give rise to multiple equilibria in the same fashion as the complex population growth functions of Section 1.2. The addition of the m function may destroy some equilibria which would exist in the absence of migration, or add new equilibria which would not exist in the absence of migration. However, since

$$d\{g + m\}/dx = dg/dx + dm/dx$$

and $dm/dx < 0$, the addition of the $m(x)$ function may make an equilibrium stable at a value of x for which $dg/dx > 0$. This could therefore be perceived as a "population stabilization" effect of migration.

More constructive analysis is possible if we add an additional assumption to model (6.15),

$$dg/dx < 0 \qquad (6.16)$$

With this assumption in hand we can immediately conclude that, if there is an $x*$ such that

$$g(x^*) + m(x^*) = 0$$

then it will be globally asymptotically stable, since

$$d\{g + m\}/dx = dg/dx + dm/dx < 0$$

and the fact that the $g + m$ function is monotone decreasing guarantees that $dx/dt > 0$ for $x < x^*$ and $dx/dt < 0$ for $x > x^*$. In addition, we can conclude that $x^* > K$ if $m(K) > 0$, and conversely $x^* < K$ if $m(K) < 0$. However, such an x^* need not exist, since the asymptotic value of $m(x)$ as x increases to ∞, say "$m(\infty)$", may be strictly positive, while the asymptotic value of $g(x)$ as x increases to ∞ could be sufficiently large that

$$g(\infty) > -m(\infty)$$

In such cases, x will increase without bound.

For concreteness, let us yet once more consider a logistic model. If both g and m functions are linear, we have

$$dx/dt = x\{r(1 - x/K) + m(1 - x/L)\} \qquad (6.17)$$

where all parameters are strictly positive. Here we take the first of the two linear functions to represent the density-dependence of net reproductive rates, while the second function represents the density-dependence of net migration rates. At equilibrium, say x^*, we have

$$x^* = (r+m)/\{r/K + m/L\}$$
$$= K\{rL + mL\}/\{rL + mK\} \qquad (6.18)$$

From equation (6.18), we can conclude that $x^* < K$ if $L < K$ while $x^* > K$ if $L > K$. In either case, it is clear that we must have x^* in the interval defined by K and L, so that it is a kind of average of the two. In effect, both reproduction and migration are functioning as mechanisms of population regulation in models of this kind, with the equilibrium population level reflecting their "average".

Two Habitats

The major lack of realism in the two basic types of migration model that we have considered so far is that the source(s) of migrants is neglected. The simplest model which adequately addresses this problem is one in which populations inhabiting two distinct habitats exchange migrants, with migration rates per individual being constant. This set of assumptions is readily embodied in the following equations:

$$dx/dt = xg(x) + c_1y - c_2x \qquad (6.19.1)$$

$$dy/dt = yh(y) - c_1y + c_2x \qquad (6.19.2)$$

where we take:

all parameters as strictly positive;
$g(0) > c_2$; $h(0) > c_1$;

$dg/dx < 0$; $dh/dy < 0$.

It is at once apparent that there are only two types of equilibria,

$$(x,y) = (0,0) \qquad (6.20)$$

or

$$(x,y) = (x*,y*) \qquad (6.21)$$

with both $x*$ and $y*$ strictly greater than zero. Equilibria of form (6.21) require

$$x*[g(x*) - c_2] = -c_1 y*$$

or

$$y* = -x*[g(x*) - c_2]/c_1 \qquad (6.22.1)$$

simultaneously with

$$y*[h(y*) - c_1] = -c_2 x*$$

or

$$x* = -y*[h(y*) - c_1]/c_2 \qquad (6.22.2)$$

This set of joint conditions in turn requires

$$g(x*) - c_2 < 0 \quad \& \quad h(y*) - c_1 < 0 \qquad (6.23)$$

and thus the existence of finite x' and y' such that $g(x') = h(y') = 0$. If such x' and y' do not exist, then the population growth trajectories will be unbounded.

The Jacobian matrix for the linearized population dynamics near the origin is

$$\begin{bmatrix} g(0) - c_2 & c_1 \\ & \\ c_2 & h(0) - c_1 \end{bmatrix} \qquad (6.24)$$

You will recall from the end of Chapter Two that local asymptotic stability requires that the trace of a matrix be less than zero. In this case the trace is given by

$$g(0) - c_2 + h(0) - c_1 > 0$$

precluding local asymptotic stability of the origin for this system. If there are any stable equilibria for model (6.19), they are in the interior of the first quadrant. In the Exercises, you are to derive the Jacobian matrix for such interior equilibria and consider the conditions for local asymptotic stability.

Let us consider the Lotka-Volterra version of model (6.19):

$$dx/dt = r_1 x - b_1 x^2 + c_1 y - c_2 x \quad (6.25.1)$$

$$dy/dt = r_2 y - b_2 y^2 - c_1 y + c_2 x \quad (6.25.2)$$

where we take all parameters to be strictly positive. If we let

$$a_1 = r_1 - c_2 > 0 \quad (6.26.1)$$

and

$$a_2 = r_2 - c_1 > 0 \quad (6.26.2)$$

then we can re-write model (6.25) as

$$dx/dt = a_1 x - b_1 x^2 + c_1 y \quad (6.27.1)$$

$$dy/dt = a_2 y - b_2 y^2 + c_2 x \quad (6.27.2)$$

At interior equilibria,

$$y^* = x^*(b_1 x^* - a_1)/c_1 \quad (6.28.1)$$

and

$$x^* = y^*(b_2 y^* - a_2)/c_2 \quad (6.28.2)$$

These equations also separately define the zero-isoclines for dx/dt and dy/dt, respectively. Jointly, they can be used to find an explicit polynomial for the equilibrium value of either variable. Take y^*, and substitute (6.28.2) into (6.28.1). We then find

$$y^* = y^*(b_2 y^* - a_2)$$
$$\times \{b_1[y^*(b_2 y^* - a_2)/c_2 - a_1\}/c_1 c_2 \quad (6.29)$$

This polynomial has an obvious root at $y^* = 0$, the root associated with the equilibrium at $(0,0)$. Once that root is eliminated from the polynomial by dividing both sides by y^*, we are left with a cubic polynomial. While it is possible to solve such cubic polynomials explicitly, that is not the avenue of analysis that we will pursue here. Instead, we note that the cubic polynomial ensures that there are at most three roots to (6.29) away from the origin, and turn to an examination of the isoclines defined by equations (6.28). These isoclines are in fact parabolas perpendicular to each other, one for y as a function of x, and one for x as a function of y. Such isoclines are plotted in Figure 6.3. Evidently, the two parabolas necessarily intersect at the origin. While these two parabolas may intersect as many as four times, when they do two of these intersection points must be outside of the admissible state-space. This is most readily seen by considering one of the parabolas at a time. Take the dx/dt = 0 parabola, (6.28.1). This parabola can also be thought of as the roots of a quadratic in x for which y is a parameter. In this light, we obtain x from

$$2x_{\pm} = a_1/b_1 \pm \sqrt{(a_1/b_1)^2 + 4yc_1/b_1} \qquad (6.30)$$

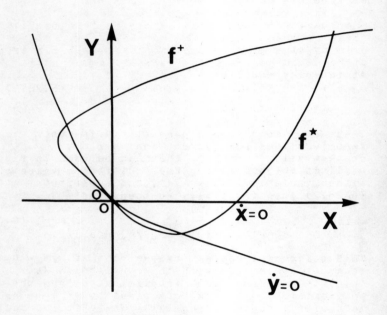

Figure 6.3. Isoclines for the Lotka-Volterra version of the "two habitat" model of population growth with migration.

For y > 0 , there are two real roots, but one is negative and the other is positive. At y = 0 , the two roots are $x_- = 0$ and $x_+ = a_1/b_1$. For

negative values of y near 0 , we retain two

roots. But if we have

$$y < -a_1^2 + 4b_1c_1 \qquad (6.31)$$

then the roots become complex, so that they are no longer admissible. In the Exercises you are to show that the RHS of (6.31) is in fact the minimum value of y for the parabola. Obviously, an exactly parallel analysis can be performed on the parabola given by (6.28.2), given the exact parallelism of form.

This functional analysis shows that there is one curve for each parabola in the interior of the first quadrant. Since the parabolas have different orientations with respect to the x-axis, they have different formulae. The expression for the $dx/dt = 0$ isocline in the first quadrant is simply

$$f^*(x) = x^2 b_1/c_1 - x a_1/c_1 \qquad (6.32.1)$$

for $x > a_1/b_1$, while the expression for the

$dy/dt = 0$ isocline is

$$f^+(x) = a_2/2b_2 + 2^{-1}\sqrt{(a_2/b_2)^2 + 4xc_2/b_2} \qquad (6.32.2)$$

for $x > 0$. We already know that $f^*(0) = 0$

while $f^+(0) > 0$. In addition,

$$df^*/dx = 2b_1x/c_1 - a_1/c_1 > 0$$

for $x > a_1/b_1$, while

$$d^2f^*/dx^2 = 2b_1/c_1 > 0$$

By contrast,

$$df^+/dx = \{(a_2/b_2)^2 + 4xc_2/b_2\}^{-1/2} 4c_1/b_2 > 0$$

for $x > 0$, while

$$d^2f^+/dx^2 = -\{(a_2/b_2)^2 + 4xc_2/b_2\}^{-3/2} 8(c_2/b_2)^2 < 0$$

for $x > 0$. From df^+/dx we can see that there

must be a finite $x` > 0$ such that $df^+/dx < 1$ for $x > x`$. Similarly, there must be a finite $x^\sim > 0$ such that $df^*/dx > 1$ for $x > x^\sim$. Using these results, you are to construct an argument in the Exercises showing that there must also be an x^*

at which f^+ and f^* intersect. This x^* will

also be unique within the interior of the first quadrant, as is also to be shown in the Exercises.

The Jacobian matrix about the interior equilibrium has the following form:

$$\begin{bmatrix} a_1 - 2b_1x^* & c_1 \\ \\ c_2 & a_2 - 2b_2y^* \end{bmatrix} \tag{6.33}$$

Since

$$a_1 - 2b_1x^* = r_1 - b_1x^* - c_2 - b_1x^*$$

$$= -c_1y^*/x^* - b_1x^*$$

we are able to re-write (6.33) as

$$\begin{bmatrix} -c_1y^*/x^* - b_1x^* & c_1 \\ \\ c_2 & -c_2x^*/y^* - b_2y^* \end{bmatrix}$$

Using the formula for the eigenvalues of a matrix, we obtain the following equation

$$(c_1y^*/x^* + b_1x^* + \lambda)(c_2x^*/y^* + b_2y^* + \lambda)$$

$$- c_1c_2 = 0 \tag{6.34}$$

This in turn becomes

$$\lambda^2 + k_1\lambda + k_2 = 0 \tag{6.35.1}$$

with

$$k_1 = c_2x^*/y^* + c_1y^*/x^* + b_1x^* + b_2y^* \tag{6.35.2}$$

and

$$k_2 = c_1b_2y^{*2}/x^* + c_2b_1x^{*2}/y^* + b_1b_2x^*y^* \tag{6.35.3}$$

In fact, characteristic equation (6.35) satisfies the Routh-Hurwitz criteria for two-dimensional systems, both k_i being strictly positive, so that

the real parts of both eigenvalues are negative, ensuring local asymptotic stability of the interior equilibrium. As yet, we do not have a Liapunov function for this system, so global stability remains uncertain.

Freedman and Waltman (1977b) have analysed a special case of this system with $c_1 = c_2 = \varepsilon$. As ε goes to ∞, x^* and y^* approach a common

limiting value of
$$[a_1 + a_2]/[b_1 + b_2]$$

An interesting feature of this equilibrium is that the sum of x* and y* can be less than the sum of their equilibrium values in the absence of migration. Thus migration can depress total species density over a range of habitats.

6.2 COMPETITION WITH MIGRATION

We will follow through the consequences of migration for populations undergoing competition using the same three basic types of model as before.

Recipient Peripheral Populations

Let x_1 and x_2 give the population densities of the two competitors in the peripheral habitat. Both of these populations are presumed to receive immigrants from a much larger conspecific population of stable population size. Consider the following model:
$$dx_1/dt = x_1 g_1(x_1) - x_1 h_1(x_2) + c_1 \qquad (6.36.1)$$

$$dx_2/dt = x_2 g_2(x_2) - x_2 h_2(x_1) + c_2 \qquad (6.36.2)$$
where

$$h_i(0) = 0 \; ; \quad dh_i/dx_j > 0 \; ;$$

$$dg_i/dx_i < 0 \; ; \quad c_i > 0 \; ;$$

there exists a $K_i > 0$ such that

$$g_i(K_i) + c_i/K_i = 0 \; .$$

This model can also be written as
$$dx_1/dt = x_1[g_1(x_1) + c_1/x_1 - h_1(x_2)] \qquad (6.37.1)$$

$$dx_2/dt = x_2[g_2(x_2) + c_2/x_2 - h_2(x_1)] \qquad (6.37.2)$$

In this form, the model bears obvious affinities with model (2.47) of Chapter Two. However, there are salient differences. One of the most important features of the present model is that there are no boundary equilibria. For x_i near 0 , we have

$$dx_i/dt \approx c_i > 0$$

Thus all equilibria are in the interior. Another important difference associated with this one is

that the zero-isoclines for the dx_i/dt extend
toward ∞ along the x_j axis. This is shown in
Figure 6.4 for the migrational analogue of Figure
2.12. As may be seen from the figure, some of the
features of Figure 2.12 are retained, such as a
region in which both populations increase in
density, another where both decrease in density,
and still others between these two where only one
species density declines while the other increases.
A major question of interest is the extent to which
the previous analysis of asymptotic trajectories
can be carried over to this model.

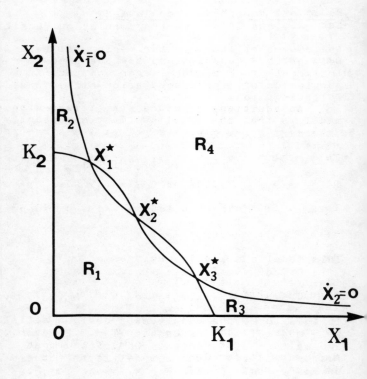

Figure 6.4. An illustrative
phase-plane portrait for model
(6.37).

The most important point is that the change in regions R_2 and R_3 compared with R_3 and R_5 of Figure 2.12 precludes the kind of topological analysis provided by Hirsch and Smale (1974, Ch. 12) in order to show that all asymptotic trajectories had to be equilibria. However, there is no point at which the analysis of local asymptotic stability of interior equilibria needs to be re-considered. Indeed, the stipulation that $dg_i/dx_i < 0$ ensures that the $f_{i,i}$ of Section 2.3 are necessarily negative. [You should check this for yourself if it is not intuitively obvious.] This reduces the requirements for local asymptotic stability to the pattern of isocline intersection: the isocline for $dx_1/dt = 0$ must cross the isocline for $dx_2/dt = 0$ from above.

Thus, in Figure 6.4, equilibria $\underline{x}_1^*$ and $\underline{x}_3^*$ are stable, while equilibrium $\underline{x}_2^*$ is not.

An additional feature that arises with this model is that the isoclines must necessarily intersect in such a way that at least one interior equilibrium will be locally asymptotically stable. This occurs because the $dx_1/dt = 0$ isocline is parallel and just to the right of the x_2 axis for large values of x_2, while it falls below the x_1 axis for $x_1 > K_1$, all of which is also true of the $dx_2/dt = 0$ isocline, with the subscripts reversed. Given continuous functions, this isocline pattern requires the intersection of the two isoclines at least once. In addition, the $dx_1/dt = 0$ isocline is above that for dx_2/dt for small values of x_1 and below it for large values, ensuring also that it must cross from above at least once. So we know that there will always be at least one asymptotically stable attractor.

Migrant Pool Model

As in Section 6.1, the migrant pool model of migration need not lead to any major change in the qualitative population dynamics of two competitor species. Let the x_i , g_i , and h_i be as before.

In addition, we have two $m_i(x)$ with the same functional features as the g_i , as in Section 6.1.

Thus we can write
$$dx_1/dt = x_1[g_1(x_1)+m_1(x_1)-h_1(x_2)] \quad (6.38.1)$$

$$dx_2/dt = x_2[g_2(x_2)+m_2(x_2)-h_2(x_1)] \quad (6.38.2)$$

In effect, we only need to define new
$$g_i'(x_i) = g_i(x_i) + m_i(x_i) \quad (6.39)$$

and invoke all the theory outlined in Section 2.3. The only consequence of migration is variation within this general class of models, as opposed to a novel dynamical structure, under "migrant pool" assumptions.

Two Habitats

It is difficult to make much headway with a generally formulated model of this type. [In the Exercises, you are to develop such a general model, by analogy with the generalized two-habitat model of Section 6.1, and analyse the dynamics near the origin.] Therefore, we will proceed directly to the Lotka-Volterra case.

Consider the following obvious extension of the classical Lotka-Volterra model of competition:
$$dx_{11}/dt = x_{11}[r_{11}-c_{11}-b_{11}x_{11}+c_{12}x_{12}/x_{11} -d_{11}x_{21}]$$

$$dx_{21}/dt = x_{21}[r_{21}-c_{21}-b_{21}x_{21}+c_{22}x_{22}/x_{21} -d_{21}x_{11}]$$

$$(6.40)$$

$$dx_{12}/dt = x_{12}[r_{12}-c_{12}-b_{12}x_{12}+c_{11}x_{11}/x_{12} -d_{12}x_{22}]$$

$$dx_{22}/dt = x_{22}[r_{22}-c_{22}-b_{22}x_{22}+c_{21}x_{21}/x_{22} -d_{22}x_{12}]$$

where we take all parameters strictly positive, with evident interpretations. Unlike the recipient peripheral population model, there can be boundary equilibria away from the origin, with $x_{11} = x_{12} = 0$ or $x_{22} = x_{21} = 0$ and the other two population densities strictly positive. In such cases, the model reduces to one of population growth with two habitats, which we have already analysed. The analysis of models of this size is not straightforward, given that the classical Lotka-Volterra assumptions are violated by the introduction of migration terms. However, it may well be a fruitful area for further work, given the importance of migration in actual ecosystems.

Levin (1974) has considered the following special form of model (6.40):

$$dx_{11}/dt = x_{11}(r-ax_{11}-bx_{21})+dx_{12}-dx_{11} \qquad (6.41.1)$$

$$dx_{21}/dt = x_{21}(r-ax_{21}-bx_{11})+dx_{22}-dx_{21} \qquad (6.41.2)$$

$$dx_{12}/dt = x_{12}(r-ax_{12}-bx_{22})+dx_{12}-dx_{12} \qquad (6.41.3)$$

$$dx_{22}/dt = x_{22}(r-ax_{22}-bx_{12})+dx_{21}-dx_{22} \qquad (6.41.4)$$

where there has been considerable notational switching. Again, all parameters are strictly positive. Here the parameter r corresponds to all four r_{ij} of model (6.40). The parameter a in fact corresponds to the four b_{ij} parameters of model (6.40), not the a_{ij} parameters. The parameter b corresponds to the d_{ij} of model (6.40). Finally, the d parameter corresponds to the c_{ij}'s of model (6.40). Levin (1974) also takes $a < b$.

With the extensive symmetries now imposed, it is apparent that the two separate competition systems which arise when $d = 0$ have the phase-plane portrait shown in Figure 6.5, where the additional parameter $K = r/a$, as is to be shown in the Exercises.

If we consider the full four-population ecosystem without migration, there are four possible configurations for locally asymptotically stable equilibria:

$$(x_{11}*,x_{12}*,x_{21}*,x_{22}*) = (K,K,0,0) \qquad (6.42.1)$$

$$(x_{11}*,x_{12}*,x_{21}*,x_{22}*) = (0,K,K,0) \qquad (6.42.2)$$

$$(x_{11}*,x_{12}*,x_{21}*,x_{22}*) = (K,0,0,K) \qquad (6.42.3)$$

$$(x_{11}*,x_{12}*,x_{21}*,x_{22}*) = (0,0,K,K) \qquad (6.42.4)$$

However, only two of these equilibrium

configurations are genuinely distinct, (6.42.1)
being analogous to (6.42.4), while (6.42.2) is
analogous to (6.42.3). These two cases correspond
to the same species predominating in both habitats
and the alternative species dominating in the two
habitats, respectively. We will discuss these two
types of cases in that order, once we have the
system Jacobian.

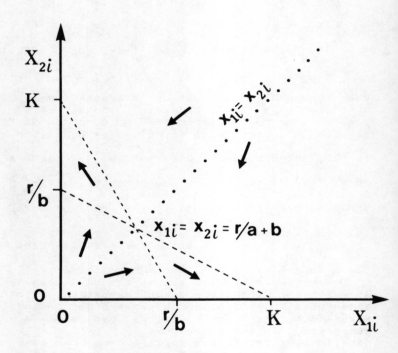

Figure 6.5. Phase-plane portrait of
the two isolated ecosystems generated
by model (6.41) when d = 0 and all
other parameters are strictly
positive.

The Jacobian is a four-by-four matrix, but it
can be thought of as a matrix composed of four
two-by-two submatrices, as follows.

$$J = \begin{bmatrix} J_1 & J_2 \\ J_3 & J_4 \end{bmatrix} \qquad (6.43.1)$$

where

$$J_1 = \begin{bmatrix} r-2ax_{11}{}^*-bx_{21}{}^*-d & d \\ d & r-2ax_{12}{}^*-bx_{22}{}^*-d \end{bmatrix} \qquad (6.43.2)$$

$$J_2 = \begin{bmatrix} -bx_{11}{}^* & 0 \\ 0 & -bx_{12}{}^* \end{bmatrix} \qquad (6.43.3)$$

$$J_3 = \begin{bmatrix} -bx_{21}{}^* & 0 \\ 0 & -bx_{22}{}^* \end{bmatrix} \qquad (6.43.4)$$

$$J_4 = \begin{bmatrix} r-2ax_{21}{}^*-bx_{11}{}^*-d & d \\ d & r-2ax_{22}{}^*-bx_{12}{}^*-d \end{bmatrix} \qquad (6.43.5)$$

Now let us consider equilibria in which the same species predominates in both habitats. We can take equilibrium form (6.42.1) without loss of generality, since equilibrium (6.42.4) is only notationally distinct. From (6.43), the Jacobian matrix for the linearized dynamics about this equilibrium is

$$\begin{bmatrix} -r-d & d & -bK & 0 \\ d & -r-d & 0 & -bK \\ 0 & 0 & r-bK-d & d \\ 0 & 0 & d & r-bK-d \end{bmatrix} \qquad (6.44)$$

As you are to show in the Exercises, this matrix

has a characteristic polynomial of form

$$[(r+d+\lambda)^2-d^2][(r-bK-d-\lambda)^2-d^2] = 0 \qquad (6.45)$$

Fortunately, this means that the eigenvalues are given by two quadratic expressions, making them readily calculable. Rather than do so, we shall use the Routh-Hurwitz criteria on each of these quadratics separately. The first of these, proceeding from left to right, can be written out as

$$\lambda^2 + 2(r+d)\lambda + r^2 + 2rd = 0 \qquad (6.46)$$

Recall that the Routh-Hurwitz criterion for a quadratic of form

$$\lambda^2 + a_1\lambda + a_2 = 0$$

is just that both a_i must be strictly positive.

In the case of quadratic (6.46), we have

$$a_1 = 2(r+d) \quad \& \quad a_2 = r^2 + 2rd$$

so that the criterion is obviously met. Before writing out the second quadratic, note that

$$r - bK = r - b(r/a) = r(1 - b/a) = L$$

Levin (1974) required $b > a$ at the outset, so we know that $L < 0$. Using L, the second quadratic can be written as

$$\lambda^2 + 2(d-L)\lambda + L^2-2Ld = 0 \qquad (6.47)$$

Now we have

$$a_1 = 2(d-L) \quad \& \quad a_2 = L^2-2Ld$$

both of which are again evidently positive. This ensures that equilibria in which the same species predominates in both habitats remain locally asymptotically stable in the face of migration. Therefore, if two habitats are sufficiently similar in both ecological parameters and ecological state, migration may have no effect.

The remaining cases involve different predominant species in the two habitats. This makes the analysis somewhat more complex. Levin (1974) has shown that for d sufficiently small, specifically d such that

$$0 \leq d < (r/d)(b-a)/(2b+a) \qquad (6.48)$$

there is a locally asymptotically stable equilibrium at

$$x_{11} = x_{22} = (r-2d-\sqrt{Y})/2a \qquad (6.49.1)$$

$$x_{12} = x_{21} = (r-2d+\sqrt{Y})/2a \qquad (6.49.2)$$
where
$$Y = (r-2d)[r-2d(b+a)/(b-a)] \qquad (6.49.3)$$

Note that as d approaches 0 arbitrarily closely, x_{11} and x_{22} approach 0 , while x_{12} and x_{21} approach $2r/2a = K$. Thus the migration parameter d smoothly perturbs the position of the locally asymptotically stable equilibrium, at least up to a threshold, giving rise to an increase in the population density of the unrepresented species in the equilibrium lacking migration. This is quite intuitively appealing, and in fact it has been made general by Levin, as will be discussed in Section 6.4. However, if d is sufficiently large, migration may have the effect of broadening interspecific competition over the two habitats such that only one species remains at equilibrium over both habitats, at least in this special case where the equations permit only one species in each habitat at equilibrium in the absence of migration. Thus, while migration may enhance the species diversity within any given habitat, it need not do so.

6.3   PREDATION WITH MIGRATION

Models of population growth or competition with migration could only reasonably assume that all populations would be equally subject to migration. In the case of predation, this assumption is not quite so self-evident. Indeed, it is often the case that predators are more vagile than their prey. Particularly when it is borne in mind that the prey may be plants and the predators herbivores, this possibility becomes evident. Accordingly, the extant models in the literature have in some cases dealt with predator-prey dynamics in which there is migration between what are two distinct habitats for the prey but only one uniform habitat for the predator. We will be treating a model of this type, along with models in which there is no such vagility distinction between predator and prey.

Recipient Peripheral Populations
As in Section 6.2, we begin with a simple perturbed version of the conventional predator-prey model of Section 3.2:

$$dx/dt \quad = \quad xg(x) - yp(x) + c_1 \qquad (6.50.1)$$

$$dy/dt \quad = \quad y[-\gamma + q(x)] + c_2 \qquad (6.50.2)$$

where all the functions and parameters have the same properties and interpretations as in model (3.25), except for the c_i , which are strictly

positive, representing the receipt of immigrants from large stable populations elsewhere. A somewhat more general version of this model has been discussed by Brauer and Soudack (1981), while a special case is discussed in Freedman (1980, pp.81-83).

For both species, their rates of change in population density approximate their respective c_i

values when their population densities are low. Thus all edges lack equilibria.

There is much information to be gained from considering the isoclines along which the rates of change are identically zero. In the case of dx/dt this occurs when

$$yp(x) \quad = \quad xg(x) + c_1$$

or

$$y \quad = \quad xg(x)/p(x) + c_1/p(x) \qquad (6.51.1)$$

Evidently, migration raises this isocline vertically, and deforms its shape. Indeed, since $p(x)$ goes to zero as x goes to zero, we must have y approaching ∞ as x goes to zero, along this isocline. The expression for the dy/dt = 0 isocline is altered even more radically. For model (3.25), it was simply x^* such that $q(x^*) = \gamma$. For model (6.50), it is

$$y \quad = \quad c_2/[\gamma - q(x)] \qquad (6.51.2)$$

which is evidently a function of x . Since $q(x)$ is an increasing positive-valued function of x , with $q(0) = 0$, we can see that y is a positive-valued and increasing function of x for $0 < x < x^*$, while y does not appear in the first quadrant for $x \geq x^*$. When $x = 0$, $y = c_2/\gamma >$

0 . Putting all this information together, immigration will give rise to the change in isocline configurations shown in Figures 6.6 and 6.7.

Most importantly, the introduction of immigration makes isocline-isocline intersection necessary, as it did in the case of the analogous model of competition with migration.

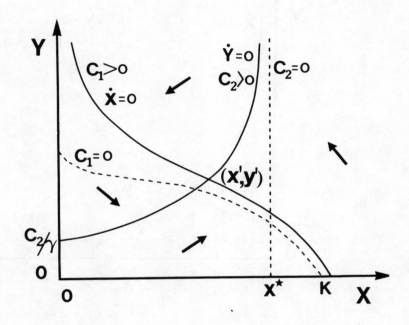

Figure 6.6. Effect of
density-independent immigration on a
predator-prey model in which
the isoclines intersect without
migration.

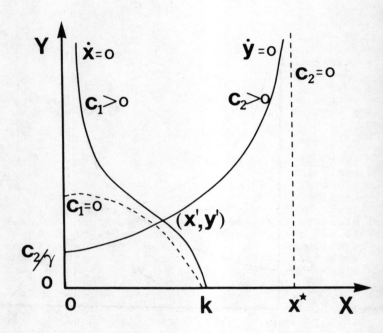

Figure 6.7. Effect of density-independent immigration on a predator-prey model in which the isoclines don't intersect without migration.

Since the immigration is not density-dependent, the Jacobian matrix for the exclusively interior equilibria is unaltered, except for the position of the equilibria. However, since the analysis of Section 3.2 relied upon the fact that $q(x*) = \gamma$, which isn't true at the equilibria with migration, we need to re-consider the analysis of that section. Explicitly, the Jacobian matrix is now

$$\begin{bmatrix} b(x',y') & -p(x') \\ y'dq(x')/dx & -\gamma + q(x') \end{bmatrix} \qquad (6.52.1)$$

where
$$b(x',y') = x'dg(x')/dx + g(x') - y'dp(x')/dx \qquad (6.52.2)$$
and y' may be obtained from the isocline equations and x' is given by the roots of
$$0 = x'g(x')/p(x') + c_1/p(x') + c_2/[\gamma-q(x')] \qquad (6.52.3)$$

It is at once obvious from the phase-plane portraits of Figures 6.6 and 6.7 that $x' < x^*$, although this in turn has complex effects on the terms defining b. It has a simple effect on the lower right-hand term of the matrix, making it strictly negative. Since the trace of matrix (6.52) must be negative for stability of the equilibrium, this effect may be thought of as enhancing stability. The characteristic equation of matrix (6.52) can be written out as
$$[\lambda - b(x',y')][\lambda + \{\gamma-q(x')\}] + p(x')y'dq(x')/dx = 0$$
If $b(x',y')$ is negative, then all the terms of this polynomial will be positive, meeting the Routh-Hurwitz criteria for local asymptotic stability.

For illustrative purposes, let us consider the Lotka-Volterra version of model (6.50):
$$dx/dt = x(a-bx-py) + c_1 \qquad (6.53.1)$$

$$dy/dt = y[-\gamma+px] + c_2 \qquad (6.53.2)$$

with all parameters strictly positive. We have isoclines
$$f_1 = a/p - bx/p + c_1/px \qquad (6.54.1)$$
and
$$f_2 = -c_2/[-\gamma + px] \qquad (6.54.2)$$

for (6.53.1) and (6.53.2), respectively. Evidently, for x in the first quadrant,

$$df_1/dx = -b/p - c_1p/x^2 < 0$$
while

$$df_2/dx = c_2 p[\gamma - px]^{-2} > 0$$

This ensures that there is a unique intersection of the two isoclines in the first quadrant. In this case, we have

$$b(x', y') = a - 2bx' - py'$$
$$= (a - bx' - py') - bx'$$
$$= -c_1/x' - bx' < 0$$

guaranteeing local asymptotic stability of the unique interior equilibrium. This doesn't quite complete the analysis, in that global convergence hasn't been demonstrated, but it does indicate that migration need not overthrow the dynamical features of the model, at least in the case of recipient peripheral populations.

Migrant Pool Model

If we make migration rates strictly functions of the prey density, with prey dispersing from habitats with high intraspecific competition and predators migrating toward areas with high prey availability, then we have the following model:

$$dx/dt = xg(x) - yp(x) + xh_1(x) \qquad (6.55.1)$$

$$dy/dt = y[-\gamma + q(x) + h_2(x)] \qquad (6.55.2)$$

where the $g(x)$, $p(x)$, and $q(x)$ functions are as before; $h_1(0) \geq 0$; $h_2(0) = 0$; $dh_1/dx < 0$;

and $dh_2/dx > 0$. As before, nothing qualitatively new arises in this model, although there will be quantitative changes in the model's dynamics. Note that if migration rates were made nonlinear functions of predator density the model structure would not be retained.

Two Habitats

Freedman and Waltman (1977b) have considered a Lotka-Volterra model in which the prey migrate between two habitats over which the predator ranges freely. In a sense, this can be thought of as perturbation to a one-predator/two-prey model, or as a model of predation on genetically distinct clones of an asexual species, where the clones are subject to mutation from one state to the other. In any case, the model is as follows:

$$dx_1/dt = \alpha_1 x_1(1-x_1/K_1) - \beta_1 x_1 y - \epsilon\mu_1 x_1 + \epsilon\mu_2 x_2 \qquad (6.56.1)$$

$$dx_2/dt = \alpha_2 x_2(1-x_2/K_2) - \beta_2 x_2 y - \epsilon\mu_2 x_2 + \epsilon\mu_1 x_1 \qquad (6.56.2)$$

$$dy/dt = y(-\gamma + \delta_1 x_1 + \delta_2 x_2) \qquad (6.56.3)$$

where all parameters are strictly positive and $\varepsilon \ll 1$. Freedman and Waltman (1977b, Theorem 4.1) show that when this system has an interior equilibrium without migration, it also has an interior equilibrium with migration and this interior point is asymptotically stable too.

6.4 ECOSYSTEMS WITH MIGRATION

Generally speaking, discussion of the effects of migration on ecosystem dynamics in deterministic models has tended to dwell upon the possibility of migration having little impact. For migrant pool models, it is apparent from preceding sections that they can be readily contrived so that there are no qualitative changes in the structure of the model equations. On the other hand, they could also be contrived so as to give rise to new terms in the ecosystem models.

Ecosystem models in which all populations receive immigrants at a fixed rate have several interesting features. The first is that all such populations cannot go extinct, and thus coexistence of all species is assured. This gives rise to an "end-run" around the problem of ecosystem persistence and model complexity. To the extent to which all species receive immigrants, they cannot go extinct, and there is then no relationship whatsoever between model complexity and stability (cf. Section 5.1). The second interesting feature of such models is that at very high population densities a fixed and low rate of immigration need not have much effect. This is readily seen by considering a single species' rate of change in population density:

$$dx_i/dt = x_i f_i(x_1, x_2, \ldots, x_n) + c_i$$

$$= x_i[f_i(x_1, \ldots, x_n) + c_i/x_i] \qquad (6.57.i)$$

Evidently, as x_i becomes arbitrarily large, the net reproductive rate of individuals of species i approaches f_i . The third interesting feature of models of this kind is that they can give rise to equilibria which did not exist before, as was illustrated in Section 6.3 for the predator-prey model. This effect is in some ways a corollary of the first effect, in that boundary equilibria are necessarily eliminated by the addition of constant-rate immigration, often with the associated creation of new interior equilibria.

While this range of effects clearly makes models of this kind mathematically interesting, they probably aren't satisfactory from the standpoint of "global theory", in that they are not appropriate to the modelling of the complete constellation of populations of a species, given their presumption that additional migrants must always come from elsewhere. The same is also true of the migrant pool models; they too have to have migrants coming from, or going to, somewhere else, making them intrinsically unfit for complete characterization of the effects of migration on a species' distribution and abundance. The only plausible models for dealing with problems of that kind are the "two habitat" models suitably generalized to m habitats. In models of this kind, the sources and destinations of all migrants are accounted for.

Levin (1974, 1976, 1978) has provided the most complete discussion of models with multiple habitats "explicitly" treated. Here we will briefly outline his main result. Let there be n species distributed over m habitats. Let the density of species i in habitat j be given by $x_{i,j}$. The local j-th habitat population density growth rate for species i, in the absence of migration, will be given by

$$f_{i,j}(x_{1,j}, x_{2,j}, \cdot \cdot \cdot, x_{n,j})$$

or just

$$f_{i,j}(\underline{x}_j)$$

where $\underline{x}_j$ is the vector of n species densities in habitat j. All $f_{i,j}$'s are assumed to be continuous and differentiable for all admissible values of the species densities. We also take $f_{i,j} \geq 0$ when $x_{i,j} = 0$. Let the migration rates $c_{i,j,k}$ be the migrational proportion between habitats k and j, such that

$$c_{i,j,k}(x_{i,j} - x_{i,k})$$

gives the net exchange between habitats j and k .

Assembling all this notation, we have

$$dx_{i,j}/dt = f_{i,j}(\underline{x}_j) + \Sigma_k c_{i,k,j}(x_{i,k}-x_{i,j})$$

$$= F_{i,j}(X,C) \qquad (6.58)$$

where

$$X = (\underline{x}_1,\underline{x}_2, \cdot \cdot \cdot ,\underline{x}_m)$$

and

$$C = (c_{1,1,1}, \cdot \cdot \cdot ,c_{1,m,1}, \cdot \cdot \cdot ,c_{n,m,m})$$

so that

$$F_{i,j}(X,\underline{0}) = f_{i,j}(\underline{x}_j)$$

$\underline{0}$ representing a vector of zero's of length nm^2 . For this system, Levin has shown that _if_ the reduced system

$$dx_{i,j}/dt = f_{i,j}(\underline{x}_j)$$

has a stable equilibrium, say X_o , _then_ the system

$$dx_{i,j}/dt = F_{i,j}(X,C)$$

has an asymptotically stable equilibrium at $X_C \geq \underline{0}$

where X_C tends to X_o as C tends to $\underline{0}$.

[Analogous results were also shown for discrete-time models by Karlin and McGregor (1972).]

What this means is that migration, if sufficiently infrequent, may only shift the asymptotic properties of ecosystems a little. This is something we have already seen in the specific low-species-number cases of preceding sections of this chapter, most explicitly at the ends of Sections 6.2 and 6.3. It suggests that the intuition that migration need not radically undermine what we know of ecosystems studied in isolation, at least theoretically, has some validity. However, it does not show that migration can not disrupt ecosystem dynamics radically. That this is possible is evident from cases like that shown in Figure 6.7, in which constant-rate immigration allows the coexistence of a predator with its prey. There are no truly general lessons; we have yet another set of theoretical problems where almost anything is possible.

6.5 EXERCISES

Elementary

1. Derive the Jacobian matrix for equilibria of form (6.21) for model (6.19).

2. Show that the RHS of (6.31) gives the minimum value of y for the parabola defined by (6.28.1).

3. Derive (6.45).

Intermediate

4. Construct an argument showing that local asymptotic stability gives rise to global asymptotic stability for model (6.1) when $dg/dx < 0$ for all $x \geq 0$.

5. Show that the f^+ and $f*$ isoclines of (6.32) must intersect in the first quadrant exactly once.

6. Develop a general model of competition with migration between two habitats and analyse its dynamics near the origin.

7. Show that when $d = 0$ the separate habitats of model (6.41) have the phase-plane portrait features shown in Figure 6.5.

Advanced

8. Analyse limit cycle existence and stability in model (6.13).

BIBLIOGRAPHY

Albrecht, F., H. Gatzke, A. Haddad, and N. Wax
    (1974) 'The dynamics of two interacting
    populations', J. Math. Anal. Appl., 46,
    658-670
Armstrong, R.A., and R.P. McGehee (1980)
    'Competitive exclusion', Amer. Nat., 115,
    151-170
Beddington, J.R., C.A. Free, and J.H. Lawton (1975)
    'Dynamic complexity in predator-prey models
    framed in difference equations', Nature
    (London), 225, 58-60
Beddington, J.A., C.A. Free, and J.H. Lawton (1976)
    'Concepts of stability and resilience in
    predator-prey models', J. Anim. Ecol., 45,
    791-816
Beddington, J. A., and P.S. Hammond (1977) 'On the
    dynamics of host-parasite-hyperparasite
    interactions', J. Anim. Ecol., 46, 811-821
Brauer, F., and A.C. Soudack (1981) 'Constant-rate
    stocking of predator-prey systems', J. Math.
    Biology, 11, 1-14
Bulmer, M.G. (1976) 'The theory of prey-predator
    oscillations' Theor. Pop. Biol., 9, 137-150
Case, T.J., and R.G. Casten (1979) 'Global
    stability and multiple domains of attraction
    in ecological systems', Amer. Nat., 113,
    705-714
Charlesworth, B. (1980) Evolution in Age-Structured
    Populations, Cambridge University Press,
    London
Comins, H.N., and M.P. Hassell (1976) 'Predation in
    multi-prey communities', J. Theor. Biol., 62,
    93-114
Coste, J., J. Peyraud, P. Coullet, and A. Chenciner
    (1978) 'About the theory of competing
    species', Theor. Pop. Biol., 14, 165-184
Cramer, N.F., and R.M. May (1972) 'Interspecific
    competition, predation, and species diversity:
    a comment', J. Theor. Biol., 34, 289-293
Cushing, J.M. (1977) Integrodifferential Equations
    and Delay Models in Population Dynamics,
    Springer-Verlag, New York
De Angelis, D.L. (1975) 'Stability and connectance
    in food web models', Ecology, 56, 238-243
Feller, W. (1968) An Introduction to Probability
    Theory and Its Applications, Vol. I, Third
    Ed., John Wiley & Sons, London

Freedman, H.I. (1980) Deterministic Mathematical
    Models in Population Ecology, Marcel Dekker,
    New York
Freedman, H.I., J.F. Addicott, and B. Rai (1983)
    'Nonobligate and obligate models of mutualism'
    in H.I. Freedman and C. Strobeck (eds.),
    Population Biology, Springer-Verlag,
    New York, pp. 349-354
Freedman, H.I., and P.M. Waltman (1977a)
    'Mathematical analysis of some three species
    food chain models', Math. Biosci., 33, 257-276
_____ (1977b) 'Mathematical models of
    species interaction with dispersal. I.
    Stability of two habitats with and without a
    predator', SIAM J. Appl. Math., 32, 631-648
Fujii, K. (1977) 'Complexity-stability relationship
    of two-prey-one-predator species systems
    model: local and global stability', J. Theor.
    Biol., 69, 613-623
Gardner, M.R., and W.R. Ashby (1970) 'Connectance
    of large dynamical (cybernetic) systems:
    critical values for stability', Nature, 228,
    784
Gause, G.F. (1934) The Struggle for Existence,
    Williams and Wilkins, Baltimore
Gause, G.F., N.P. Smaragdova, and A.A. Witt (1936)
    'Further studies of interaction between
    predators and prey', J. Anim. Ecol., 5, 1-18
Gilpin, M.E. (1975) 'Limit cycles in competition
    communities', Amer. Nat., 109, 51-60
_____ (1979) 'Spiral chaos in a
    predator-prey model', Amer. Nat., 113, 306-308
Goh, B.S. (1977) 'Global stability in many species
    systems', Amer. Nat., 111, 135-143
_____ (1980) Management and Analysis of
    Biological Populations, Elsevier, New York
Haigh, J., and J. Maynard Smith (1972) 'Can there
    be more predators than prey?', Theor. Pop.
    Biol., 3, 290-299
Hassell, M.P. (1978) The Dynamics of Arthopod
    Predator-Prey Systems, Princeton University
    Press, Princeton, N.J.
_____ (1979) 'The dynamics of predator-prey
    interactions: polyphagous predators, competing
    predators and hyperparasitoids' in R.M.
    Anderson, B.D. Turner, and L.R. Taylor (eds.),
    Population Dynamics, Oxford University Press,
    London, pp.283-306
Hassell, M.P., and H.N. Comins (1976) 'Discrete-
    time models for two-species competition',
    Theor. Pop. Biol., 9, 202-221

Hassell, M.P., and R.M. May (1973) 'Stability in insect host-parasite models', _J. Anim. Ecol._, 42, 693-726

Hirsch, M.W., and S. Smale (1974) _Differential Equations, Dynamical Systems, and Linear Algebra_, Academic Press, New York

Holling, C.S. (1973) 'Resilience and stability of ecological systems', _Ann. Rev. Ecol. & Syst._, 4, 1-23

Holt, R.D. (1983a) 'Immigration and the dynamics of peripheral populations' in _Advances in Herpetology and Evolutionary Biology_, Museum of Comparative Zoology, Cambridge, Mass., pp. 680-694

_____ (1983b) 'Models for peripheral populations: The role of immigration' in H.I. Freedman and C. Strobeck (eds.), _Population Biology_, Springer-Verlag, New York, pp. 25-32

Hoppensteadt, F.C. (1976) _Mathematical Methods of Population Biology_, Courant Institute, New York

Hutchinson, G.E. (1948) 'Circular causal systems in ecology', _Ann. N.Y. Acad. Sci._, 50, 221-246

Karlin, S., and J. McGregor (1972) 'Polymorphisms for genetic and ecological systems with weak coupling', _Theor. Pop. Biol._, 26, 210-238

Kolmogorov, A. (1936) 'Sulla teoria di Volterra della lotta per l'esistenza', _Gi. Inst. Ital. Attuari_, 7, 74-80

Krebs, C.J. (1985) _Ecology, The Experimental Study of Distribution and Abundance_, Third Ed., Harper and Row, New York

León, J.A. (1975) 'Limit cycles in populations with separate generations', _J. Theor. Biol._, 49, 241-244

Leslie, P.H. (1945) 'On the use of matrices in certain population mathematics', _Biometrika_, 33, 183-212

_____ (1948) 'Some further notes on the use of matrices in population mathematics', _Biometrika_, 35, 213-245

Levin, S.A. (1974) 'Dispersion and population interactions', _Amer. Nat._, 108, 207-228

_____ (1976) 'Spatial patterning and the structure of ecological communities' in S.A. Levin (ed.), _Some Mathematical Questions in Biology_, Amer. Math. Soc., Providence, R.I., pp.1-36

_____ (1978) 'Population models and community structure in heterogeneous environments' in

S.A. Levin (ed.), Studies in Mathematical Biology, Vo. 16, Part II: Populations and Communities, Math. Assoc. Amer., Washington, D.C., 439-476

Levins, R. (1968) Evolution in Changing Environments, Princeton University Press, Princeton, N.J.

Lewis, E.G. (1942) 'On the generation and growth of a population', Sankhya, 6, 93-96

Li, T-Y., and J.A. Yorke (1975) 'Period three implies chaos', Amer. Math. Mon., 82, 985-992

MacDonald, N. (1978) Time Lags in Biological Models, Springer-Verlag, Berlin

Malthus, T.R. (1798 [1926]) An Essay on the Principles of Population as it Affects the Future Improvement of Society with Remarks on the Speculations of Mr. Godwin, M. Condorcet and Other Writers, Facsimilie Ed., Macmillan, London

Marsden, J.E., and M. McCracken (1976) The Hopf Bifurcation and Its Applications, Springer-Verlag, New York

May, R.M. (1971) 'Stability in multi-species community models', Math. Biosci., 12, 59-79

_____ (1972) 'Will a large complex system be stable?', Nature, 238, 413-414

_____ (1973) 'Time-delay versus stability in population models with two and three trophic levels', Ecology, 54, 315-325

_____ (1974) Stability and Complexity in Model Ecosystems, Second Ed., Princeton University Press, Princeton, N.J.

_____ (1977) 'Thresholds and breakpoints in ecosystems with a multiplicity of stable states', Nature, 269, 471-477

_____ (1978) 'Host-parasitoid systems in patchy environments: a phenomenological model', J. Anim. Ecol., 47, 833-843

_____ (1979) 'The structure and dynamics of ecological communities' in R.M. Anderson, B.D. Turner, and L.R. Taylor (eds.), Population Dynamics, Blackwell Scientific Publications, Oxford, U.K., pp.385-407

_____ (1981) 'Models for Two Interacting Populations' in R.M. May (ed.), Theoretical Ecology, Principles and Applications, Second Ed., Blackwell Scientific Publications, Oxford, U.K., pp.78-104

May, R.M., and M.P. Hassell (1981) 'The dynamics of multiparasitoid-host interactions', Amer. Nat., 117, 234-261

Bibliography

May, R.M., and W.J. Leonard (1975) 'Nonlinear aspects of competition between three species', SIAM J. Appl. Math., 29, 243-253

May, R.M., and G.F. Oster (1976) 'Bifurcations and dynamic complexity in simple ecological models', Amer. Nat., 110, 573-599

Maynard Smith, J. (1968) Mathematical Ideas in Biology, Cambridge University Press, London
_____ (1974) Models in Ecology, Cambridge University Press, London

McGehee, R., and R.A. Armstrong (1977) 'Some mathematical problems concerning the ecological principle of competitive exclusion', J. Differ. Equations, 23, 30-52

McMurtie, R.E. (1975) 'Determinants of stability of large randomly connected systems', J. Theor. Biol., 50, 1-11

Nagylaki, T. (1977) Selection in One- and Two-Locus Systems, Springer-Verlag, New York

Nicholson, A.J. (1933) 'The balance of animal populations', J. Anim. Ecol., 2, 553-565

Nicholson, A.J., and V.A. Bailey (1935) 'The balance of animal populations', Proc. Zool. Soc. Lond. Part I, 1935, 551-598

Pielou, E.C. (1977) An Introduction to Mathematical Ecology, Second Ed., Wiley-Interscience, New York

Pimm, S.L., and J.H. Lawton (1980) 'Are food webs divided into compartments?', J. Anim. Ecol., 49, 879-898

Pullman, N.J. (1976) Matrix Theory and Its Applications: Selected Topics, Marcel Dekker, New York

Rescigno, A., and K.G. Jones (1972) 'The struggle for life. III: A predator-prey chain,' Bull. Math. Biophys., 34, 521-532

Rescigno, A., and I.W. Richardson (1967) 'The struggle for life. I: Two species', Bull. Math. Biophys., 29, 377-388

Rosenzweig, M.L. (1973) 'Exploitation in three trophic levels', Amer. Nat., 47, 209-223

Rosenzweig, M.L., and R.H. MacArthur (1963) 'Graphical representation and stability conditions of predator-prey interactions', Amer. Nat., 47, 209-233

Roughgarden, J. (1979) Theory of Population Genetics and Evolutionary Ecology: An Introduction, Collier MacMillan, London

Samuelson, P.A. (1967) 'A universal cycle?', Oper. Res., 3, 307-320

Scudo, F.M., and J.R. Ziegler (1978) The Golden Age

of Theoretical Ecology: 1923-1940,
Springer-Verlag, New York
Takeuchi, Y., and N. Adachi (1983) 'Oscillations in
prey-predator Volterra models' in H.I.
Freedman and C. Strobeck (eds.) Population
Biology, Springer-Verlag, New York, pp.320-326
Taylor, L.R., I.P. Woiwod, and J.N. Perry (1979)
'The negative binomial as a dynamic ecological
model for aggregation, and the density
dependence of k', J. Anim. Ecol., 48, 289-304
Tuljapurkar, S.D., and J.S. Semura (1979) 'Liapunov
functions: geometry and stability', J. Math.
Biol., 8, 25-32
Vance, R.R. (1978) 'Predation and resource
partitioning in one predator-two prey model
communities', Amer. Nat., 112, 797-813
Wangersky, P.J. (1978) 'Lotka-Volterra population
models', Ann. Rev. Ecol. Syst., 9, 189-218
Wangersky, P.J., and W.J. Cunningham (1956) 'On
time lags in equations of growth', Proc. Nat.
Acad. Sci. USA, 42, 699-702
_____ (1957a) 'Time lag in
population models', Cold Spring Harbor Symp.
Quant. Biol., 22, 329-338
_____ (1957b) 'Time lag in
prey-predator population models', Ecology, 38,
136-139
Wynne-Edwards, V.C. (1962) Animal Dispersion in
Relation to Social Behaviour, Oliver and Boyd,
Edinburgh